SpringerBriefs in Mathematics

T0202857

SpringerBriefs in Mathematics showcases expositions in all areas of mathematics and applied mathematics. Manuscripts presenting new results or a single new result in a classical field, new field, or an emerging topic, applications, or bridges between new results and already published works, are encouraged. The series is intended for mathematicians and applied mathematicians.

For further volumes:
http://www.springer.com/series/10030

BCAM SpringerBriefs

BCAM *SpringerBriefs* aims to publish contributions in the following disciplines: Applied Mathematics, Finance, Statistics and Computer Science. BCAM has appointed an Editorial Board that will evaluate and review proposals.

Typical topics include: a timely report of state-of-the-art analytical techniques, bridge between new research results published in journal articles and a contextual literature review, a snapshot of a hot or emerging topic, a presentation of core concepts that students must understand in order to make independent contributions.

Please submit your proposal to the Editorial Board or to Francesca Bonadei, Executive Editor Mathematics, Statistics, and Engineering: francesca.bonadei@springer.com

Ludovic Rifford

Sub-Riemannian Geometry and Optimal Transport

basque center for applied mathematics

Ludovic Rifford
Laboratoire J.A. Dieudonné
Université Nice Sophia Antipolis
Nice
France

ISSN 2191-8198   ISSN 2191-8201 (electronic)
ISBN 978-3-319-04803-1   ISBN 978-3-319-04804-8 (eBook)
DOI 10.1007/978-3-319-04804-8
Springer Cham Heidelberg New York Dordrecht London

Library of Congress Control Number: 2014933272

Printed on acid-free paper

Springer is part of Springer Science+Business Media (www.springer.com)

Preface

The main goal of these lectures is to give an introduction to sub-Riemannian geometry and optimal transport, and to present some of the recent progress in these two fields. This set of notes is divided into three chapters and two appendices. Chapter 1 is concerned with the notions of totally nonholonomic distributions and sub-Riemannian structures. The concepts of End-Point mappings and singular horizontal paths which play a major role through these lectures are introduced here. Chapter 2 deals with sub-Riemannian geodesics. We study first- and second-order variations of the End-Point mapping to derive necessary and sufficient conditions for an horizontal path to be minimizing. In Chap. 3, we study the Monge problem for sub-Riemannian quadratic costs. We give a crash-course in optimal transport theory and explain how the sub-TWIST condition together with the Lipschitz regularity of a "variational" cost implies the well-posedness of Monge's problem. Then, we study the fine regularity properties of sub-Riemannian distances to obtain existence and uniqueness of optimal transport maps in the sub-Riemannian context. We recall basic facts on ordinary differential equations in Appendix A and less classical results of differential calculus in normed vector spaces in Appendix B. The latter plays a key role in Chap. 2.

The reader of these notes should be familiar with the basics in differential geometry and measure theory. For further reading, we strongly encourage the reader to look at other texts in sub-Riemannian geometry and optimal transport. Multiple viewpoints always lead to deeper understanding and may open new directions for research. Among them, we may suggest the textbooks by Montgomery [2], Agrachev, Barilari and Boscain [1], and Villani [3].

This set of notes grew from a series of lectures that I gave during a CIMPA school in Beyrouth, Lebanon, on the invitation of Fernand Pelletier. I take the opportunity of this preface to warmly thank Ali Fardoun, Mohamad Mehdi, and Fernand Pelletier who organized the school, Ahmed El Soufi for his support and friendship, and through him the "Centre International de Mathématiques Pures et Appliquées." My gratitude goes also to all faculties and students who attended this sub-Riemannian CIMPA school in making it a success.

Nice, June 2013
Ludovic Rifford

References

1. Agrachev, A., Barilari, D., Boscain, U.: Introduction to Riemannian and sub-Riemannian geometry. To appear
2. Montgomery, R.: A tour of subriemannian geometries, their geodesics and applications. In: Mathematical Surveys and Monographs, vol. 91. American Mathematical Society, Providence, RI (2002)
3. Villani, C.: Optimal transport, Old and new. Springer-Verlag, Heidelberg (2008)

Contents

Chapter 1
Sub-Riemannian Structures

Throughout all the chapter, M denotes a smooth connected manifold without boundary of dimension $n \geq 2$.

1.1 Totally Nonholonomic Distributions

Distributions. A smooth *distribution* Δ of rank $m \leq n$ ($m \geq 1$) on M is a rank m subbundle of the tangent bundle TM, that is a smooth map that assigns to each point x of M a linear subspace $\Delta(x)$ of the tangent space $T_x M$ of dimension m. In other terms, for every $x \in M$, there are an open neighborhood $\mathcal{V}_x$ of x in M and m smooth vector fields $X_x^1, \ldots, X_x^m$ linearly independent on $\mathcal{V}_x$ such that

$$\Delta(y) = \mathrm{Span}\left\{X_x^1(y), \ldots, X_x^m(y)\right\} \qquad \forall y \in \mathcal{V}_x.$$

Such a family of smooth vector fields is called a *local frame* in $\mathcal{V}_x$ for the distribution Δ. All the distributions which will be considered later will be smooth with constant rank $m \in [1, n]$. Thus, from now on, "distribution" always means "smooth distribution with constant rank". A co-rank k distribution on M is a distribution of rank $m = n - k$ and any smooth vector field X on M such that $X(x) \in \Delta(x)$ for any $x \in M$ is called a section of Δ.

Example 1.1 We call *trivial distribution* on M the rank n distribution Δ defined by $\Delta(x) = T_x M$ for all $x \in M$. For topological reasons, such a distribution may not admit non-vanishing sections (for example, by the hairy ball theorem, there is no non-vanishing continuous vector fields on any even dimensional sphere).

Example 1.2 In $\mathbb{R}^3$ with coordinates (x, y, z), the distribution Δ generated by the vector fields X and Y, that is

$$\Delta(x, y, z) = \mathrm{Span}\left\{X(x, y, z), Y(x, y, z)\right\} \qquad \forall (x, y, z) \in \mathbb{R}^3,$$

L. Rifford, *Sub-Riemannian Geometry and Optimal Transport*,
SpringerBriefs in Mathematics, DOI: 10.1007/978-3-319-04804-8_1,
© The Author(s) 2014

with

$$X = \partial_x - \frac{y}{2}\partial_z \text{ and } Y = \partial_y + \frac{x}{2}\partial_z,$$

is a rank 2 (or co-rank 1) distribution on $\mathbb{R}^3$.

Example 1.3 More generally, if $x = (x_1, \ldots, x_n, y_1, \ldots, y_n, z)$ denotes the coordinates in $\mathbb{R}^{2n+1}$ and the $2n$ smooth vector fields $X^1, \ldots, X^n, Y^1, \ldots, Y^n$ are defined by

$$X^i = \partial_{x_i} - \frac{y_i}{2}\partial_z, \quad Y^i = \partial_{y_i} + \frac{x_i}{2}\partial_z \quad \forall i = 1, \ldots, n,$$

then the distribution generated by the above vector fields is a co-rank 1 distribution on $\mathbb{R}^{2n+1}$.

Example 1.4 Let α be a smooth non-degenerate 1-form on M, that is a 1-form which does not vanish ($\alpha_x \neq 0$ for any $x \in M$). The distribution Δ defined as

$$\Delta(x) = \text{Ker}(\alpha_x) \quad \forall x \in M,$$

is a co-rank 1 distribution on M.

We say that a given distribution Δ on M admits a *global frame* if there are m smooth vector fields $X^1, \ldots, X^m$ on M such that

$$\Delta(x) = \text{Span}\left\{X^1(x), \ldots, X^m(x)\right\} \quad \forall x \in M.$$

In general, distributions do not admit global frames (see Example 1.1). It is worth noticing that in the particular case of $\mathbb{R}^n$ all distributions are trivial.

Proposition 1.1 *Any distribution on $\mathbb{R}^n$ admits a global frame.*

Proof Let us first show how to construct a non-vanishing section of a given distribution on $\mathbb{R}^n$.

Lemma 1.2 *Let Δ be a distribution of rank m on $\mathbb{R}^n$. Then there is a non-vanishing smooth vector field X such that $X(x) \in \Delta(x)$, for any $x \in \mathbb{R}^n$.*

Proof (Proof of Lemma 1.2) Define the multivalued mapping $\delta : \mathbb{R}^n \to 2^{\mathbb{R}^n}$ by

$$\delta(x) = \left\{v \in \Delta(x) \mid |v| = 1\right\} \quad \forall x \in \mathbb{R}^n.$$

By construction, δ is locally Lipschitz with respect to the Hausdorff distance on compact subsets of $\mathbb{R}^n$. By compactness of $\bar{B}(0_n, 2)$, there is $\varepsilon \in (0, 1)$ such that for any $x, y \in \bar{B}(0_n, 2)$ with $|x - y| < \varepsilon$, and any $v \in \delta(x)$, there is $w \in \delta(y)$ such that $|v - w| < 1$. Let $N \geq 2$ be an integer such that the increasing sequence of balls $\mathscr{B}_1, \ldots, \mathscr{B}_N$ defined by

$$\mathscr{B}_i = B(0_n, i\varepsilon) \qquad \forall i = 1, \ldots, N,$$

satisfies $\bar{B}(0_n, 1) \subset \mathscr{B}_N$. For every $x \in \mathbb{R}^n$, we denote by $\mathrm{Proj}_{\delta(x)}$ the projection onto the $(m-1)$-dimensional sphere $\delta(x)$. Note that the mapping $\mathrm{Proj}_{\delta(x)}$ is well-defined and "smooth" on the open set

$$\mathcal{O}_x = \left\{ w \in \mathbb{R}^n \mid \langle v, w \rangle \neq 0 \text{ for some } v \in \delta(x) \right\}.$$

For every $i \in \{1, \ldots, N-1\}$, consider a smooth mapping $P_i : \mathscr{B}_{i+1} \to \mathscr{B}_i$ such that

$$|P_i(x) - x| < \varepsilon \qquad \forall x \in \mathscr{B}_{i+1}, \tag{1.1}$$

and let $\bar{w} \in \delta(0)$ be fixed. We define the vector field $X : \bar{B}(0_n, 1) \to \mathbb{R}^n$ as follows: We first set

$$X_1(x) = \mathrm{Proj}_{\delta(x)}(\bar{w}) \qquad \forall x \in \mathscr{B}_1.$$

Then, given $X_i : \mathscr{B}_i \to \mathbb{R}^n$, we define $X_{i+1} : \mathscr{B}_{i+1} \to \mathbb{R}^n$ as

$$X_{i+1}(x) = \mathrm{Proj}_{\delta(x)}\left(X_i(P_i(x)) \right) \qquad \forall x \in \mathscr{B}_{i+1}.$$

By construction (by (1.1) and the definition of ε), $X_i(P_i(x))$ belongs to $\mathcal{O}_x$ for any $x \in \mathscr{B}_{i+1}$. In conclusion, $X = X_N$ is smooth on $\bar{B}(0_n, 1)$ and satisfies $0_n \neq X(x) \in \delta(x)$ for any $x \in B(0_n, 1)$. Repeating the construction on the annuli $B(0_n, 2) \setminus B(0_n, 1), B(0_n, 3) \setminus B(0_n, 2), \ldots$, we obtain a non-vanishing section of Δ on $\mathbb{R}^n$. $\qquad\square$

We now prove Proposition 1.1 by induction on m. Let Δ be a rank $(m+1)$ distribution on $\mathbb{R}^n$. By Lemma 1.2, it admits a non-vanishing section X on $\mathbb{R}^n$. The multivalued mapping $\tilde{\Delta} : \mathbb{R}^n \to 2^{\mathbb{R}^n}$ defined by

$$\tilde{\Delta}(x) = \Delta(x) \cap \left\{ X(x) \right\}^{\perp} \qquad \forall x \in \mathbb{R}^n,$$

is a smooth rank m distribution (here $\{X(x)\}^{\perp}$ denotes the space which is orthogonal to $X(x)$ with respect to the Euclidean scalar product). Thus by induction, there are smooth vector fields $X^1, \ldots, X^m$ which generate $\tilde{\Delta}$ on $\mathbb{R}^n$. The family $\{X^1, \ldots, X^m, X\}$ is a global frame for Δ. $\qquad\square$

A finite family of smooth vector fields $\{X^1, \ldots, X^k\}$ is called a *generating family* for Δ on M if there holds

$$\Delta(x) = \mathrm{Span}\left\{X^1(x), \ldots, X^k(x)\right\} \qquad \forall x \in M.$$

Since vector fields of a generating family are not necessarily linearly independent, any distribution can be represented by a generating family.

Proposition 1.3 *Let Δ be a distribution of rank $m \leq n$ on M. Then there are $k = m(n + 1)$ smooth vector fields $X^1, \ldots, X^k$ such that $\{X^1, \ldots, X^k\}$ is a generating family for Δ.*

Proof By definition, for every $x \in M$, there is an open neighborhood $\mathcal{V}_x$ of x in M and m smooth vector fields $X_x^1, \ldots, X_x^m$ linearly independent on $\mathcal{V}_x$ such that

$$\Delta(y) = \mathrm{Span}\left\{X_x^1(y), \ldots, X_x^m(y)\right\} \qquad \forall y \in \mathcal{V}_x.$$

Since M is paracompact, there is a locally finite covering $\mathcal{V} = \{\mathcal{V}_i\}_{i \in I}$ where each open set $\mathcal{V}_i$ equals $\mathcal{V}_{x_i}$ for some $x_i \in M$.

Lemma 1.4 *There are a locally finite open covering $\{\mathcal{U}_j\}_{j \in J}$ of M and a partition $\cup_{l=1}^{n+1} J_l$ of J such that the following properties are satisfied:*

(a) For every $j \in J$, there is $i = i(j) \in I$ such that $\mathcal{U}_j \subset \mathcal{V}_i$.
(b) For every $l \in \{1, \ldots, n+1\}$ and any $j \neq j' \in J_l$, $\mathcal{U}_j \cap \mathcal{U}_{j'} = \emptyset$.

Proof (Proof of Lemma 1.4) Recall that every smooth manifold is triangulable. Let $\mathcal{T} = \{\mathcal{T}_t\}_{t \in T}$ be a triangulation of M that refines the covering $\{\mathcal{V}_i\}_{i \in I}$, in the sense that the closure of each face F of $\mathcal{T}$ is a subset of some $\mathcal{V}_i$. For every $\alpha \in \{0, \ldots, n\}$, denote by $\mathcal{T}^\alpha = \{\mathcal{T}_t^\alpha\}_{t \in T_\alpha}$ the family of α-dimensional faces in $\mathcal{T}$. For every $\alpha \in \{0, \ldots, n\}$, we can construct easily a collection of open sets $\mathcal{W}^\alpha = \{\mathcal{W}_s^\alpha\}_{s \in S_\alpha}$ satisfying the following properties:

- $\mathcal{W}^\alpha$ is a refinement of $\{\mathcal{V}_i\}_{i \in I}$;
- $\cup_{t \in T_\alpha} \mathcal{T}_t^\alpha \subset \cup_{s \in S_\alpha} \mathcal{W}_s^\alpha$;
- each $\mathcal{W}_s^\alpha$ is an open neighborhood of some α-dimensional face of $\mathcal{T}^\alpha$;
- for any $s \neq s' \in S_\alpha$, $\mathcal{W}_s^\alpha \cap \mathcal{W}_{s'}^\alpha = \emptyset$;
- for any $s \neq s' \in S_0$, $\overline{\mathcal{W}_s^\alpha} \cap \overline{\mathcal{W}_{s'}^\alpha} = \emptyset$;
- for any $\alpha \in \{1, \ldots, n\}$ and any $s \neq s' \in S_\alpha$, $\overline{\mathcal{W}_s^\alpha} \cap \overline{\mathcal{W}_{s'}^\alpha} \subset \cup_{t \in T_{\alpha-1}} \mathcal{T}_t^{\alpha-1}$.

For that, it suffices to proceed by induction on α and to make use of the properties of a triangulation. We conclude easily.   □

Let us now show how to construct for every $r \in \{1, \ldots, m\}$ a family of sections $\{X_1^j, \ldots, X_{n+1}^j \mid 1 \leq j \leq r\}$ of Δ such that $\mathrm{Span}\{X_l^j(x) \mid 1 \leq j \leq r, 1 \leq l \leq n+1\}$ has dimension $\geq r$ for any $x \in M$. We proceed by induction on r.

First, for each $l \in \{1, \ldots, n+1\}$ and each $j \in J_l$, there is $i = i(j) \in I$ such that $\mathcal{U}_j \subset \mathcal{V}_i = \mathcal{V}_{x_i}$. Modifying $X_i^1 = X_{x_i}^1$ outside $\mathcal{U}_j$ if necessary, we may assume that X_i^1 is defined on M, does not vanish on $\mathcal{U}_j$, and vanishes outside $\mathcal{U}_j$. Define $X_1^1, \ldots, X_{n+1}^1$ by

$$X_l^1 = \sum_{j \in J_l} X_{i(j)}^1 \qquad \forall l = 1, \ldots, n+1.$$

By construction (Lemma 1.4 (b)), the interior of the supports of the $X_{i(j)}^1$'s are always disjoint. Therefore, each X_l^1 is a non-vanishing section of Δ on $\cup_{j \in J_l} \mathcal{U}_j$. This shows that $\mathrm{Span}\{X_l^1(x) \mid 1 \leq l \leq n+1\}$ has dimension ≥ 1 for any $x \in M$.

Assume now that we have constructed a family of smooth vector fields $\{X_i^j, \mid 1 \leq j \leq r, 1 \leq i \leq n+1\}$ such that

$$\mathrm{Span}\Big\{X_l^j(x) \mid 1 \leq j \leq r, 1 \leq l \leq n+1\Big\}$$

has dimension $\geq r$ for any $x \in M$ (with $r < m$). For every $j \in J$, there is $s = s(j) \in \{1, \ldots, m\}$ such that

$$\mathrm{Span}\Big\{X_{x_i(j)}^s(x), X_l^j(x) \mid 1 \leq j \leq r, 1 \leq l \leq n+1\Big\}$$

has dimension $\geq r+1$ for any $x \in \mathcal{U}_j$. Define $X_1^{r+1}, \ldots, X_{n+1}^{r+1}$ by

$$X_l^{r+1} = \sum_{j \in J_l} X_{i(j)}^{s(j)} \qquad \forall l = 1, \ldots, n+1.$$

We leave the reader to check that by construction (modifying the $X_{x_i(j)}^{s(j)}$'s if necessary as above), the vector space

$$\mathrm{Span}\Big\{X_l^j(x) \mid 1 \leq j \leq r+1, 1 \leq l \leq n+1\Big\}$$

has dimension $\geq r+1$ for any $x \in M$. The proof is complete. $\qquad\qquad\qquad\square$

The Hörmander condition. Recall that for any smooth vector fields X, Y on M given by

$$X(x) = \sum_{i=1}^n a_i(x) \partial_{x_i}, \quad Y(x) = \sum_{i=1}^n b_i(x) \partial_{x_i},$$

in local coordinates $x = (x_1, \ldots, x_n)$, the *Lie bracket* $[X, Y]$ is the smooth vector field defined as

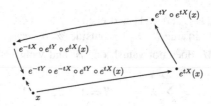

Fig. 1.1 The Lie bracket $[X, Y](x)$ measures the extent to which X and Y do not commute from x

$$[X, Y](x) = \sum_{i=1}^{n} c_i(x)\partial_{x_i},$$

where $c_1, \ldots, c_n$ are the smooth scalar function given by

$$c_i = \sum_{j=1}^{n} \left(\partial_{x_j} b_i\right) a_j - \left(\partial_{x_j} a_i\right) b_j \quad \forall i = 1, \ldots, n.$$

For the upcoming controllability results, it is important to keep in mind the following dynamical characterization of the Lie bracket (see Fig. 1.1).

Proposition 1.5 *Let X, Y be two smooth vector fields in an neighborhood of $x \in \mathbb{R}^n$. Then we have*

$$[X, Y](x) = D_x Y \cdot X(x) - D_x X \cdot Y(x) = \lim_{t \to 0} \frac{\left(e^{-tY} \circ e^{-tX} \circ e^{tY} \circ e^{tX}\right)(x) - x}{t^2}, \quad (1.2)$$

where e^{tX} and e^{tY} denote respectively the flows of X and Y.

Proof All the functions appearing in the proof will be defined locally for t close to 0 and/or in a neighborhood of x. Define the smooth function h_4 by

$$h_4(t) := \left(e^{-tY} \circ e^{-tX} \circ e^{tY} \circ e^{tX}\right)(x) \quad \forall t.$$

We have $h_4'(0) = 0$. As a matter of fact, we have for any t,

$$h_4'(t) = -Y(h_4(t)) + \left(\frac{\partial e^{-tY}}{\partial x}\right)_{(t, h_3(t))} \cdot h_3'(t)$$

where h_3 is defined by $h_3(t) := \left(e^{-tX} \circ e^{tY} \circ e^{tX}\right)(x)$. Then we have

$$h_3'(t) = -X(h_3(t)) + \left(\frac{\partial e^{-tX}}{\partial x}\right)_{(t, h_2(t))} \cdot h_2'(t),$$

where $h_2(t) := \left(e^{tY} \circ e^{tX}\right)(x)$ and

$$h_2'(t) = Y(h_2(t)) + \left(\frac{\partial e^{tY}}{\partial x}\right)_{(t,h_1(t))} \cdot h_1'(t),$$

with $h_1(t) := e^{tX}(x)$ and $h_1'(t) = X(e^{tX}(x))$. Since partial derivatives of the form $\frac{\partial e^{tX}}{\partial x}$ at $t = 0$ are equal to Id, we get $h_1'(0) = X(x)$, $h_2'(0) = X(x) + Y(x)$, $h_3'(0) = Y(x)$ and $h_4'(0) = 0$. Therefore, the left-hand side of (1.2) is equal to $\frac{1}{2}h_4''(0)$. By derivating the above formulas, we get

$$\begin{cases} h_1''(0) = D_{h_1(0)}X \cdot h_1'(0) = D_x X \cdot X(x), \\ h_2''(0) = D_{h_2(0)}Y \cdot h_2'(0) + \left[\frac{d}{dt}\left[\left(\frac{\partial e^{tY}}{\partial x}\right)_{(t,h_1(t))} \cdot h_1'(t)\right]\right]_{t=0}. \end{cases}$$

But $D_{h_2(0)}Y \cdot h_2'(0) = D_x Y \cdot (X(x) + Y(x))$ and

$$\left[\frac{d}{dt}\left[\left(\frac{\partial e^{tY}}{\partial x}\right)_{(t,h_1(t))} \cdot h_1'(t)\right]\right]_{t=0}$$

$$= \left[\frac{d}{dt}\left(\frac{\partial e^{tY}}{\partial x}\right)_{(t,h_1(t))}\right]_{t=0} \cdot h_1'(0) + \left(\frac{\partial e^{tY}}{\partial x}\right)_{(0,h_1(0))} \cdot h_1''(0)$$

$$= \left[\left(\frac{\partial^2 e^{tY}}{\partial t \partial x}\right)_{(0,x)} + \left(\frac{\partial^2 e^{tY}}{\partial x^2}\right)_{(0,x)} \cdot h_1'(0)\right] \cdot X(x) + D_x X \cdot X(x)$$

$$= \left(\frac{\partial}{\partial x}\left(\frac{\partial e^{tY}}{\partial t}\right)\right)_{(0,x)} \cdot X(x) + D_x X \cdot X(x) = D_x Y \cdot X(x) + D_x X \cdot X(x).$$

We infer that $h_2''(0) = D_x Y \cdot (2X(x) + Y(x)) + D_x X \cdot X(x)$. In the same way, we have

$$h_3''(0) = -D_{h_3(0)}X \cdot h_3'(0) + \left[\frac{d}{dt}\left[\left(\frac{\partial e^{-tX}}{\partial x}\right)_{(t,h_2(t))} \cdot h_2'(t)\right]\right]_{t=0},$$

with $-D_{h_3(0)}X \cdot h_3'(0) = -D_x X \cdot Y(x)$ and

$$\left[\frac{d}{dt}\left[\left(\frac{\partial e^{-tX}}{\partial x}\right)_{(t,h_2(t))} \cdot h_2'(t)\right]\right]_{t=0} = -D_x X \cdot Y(x) + D_x Y \cdot (2X(x) + Y(x)),$$

which implies $h_3''(0) = -2D_x X \cdot Y(x) + D_x Y \cdot (2X(x) + Y(x))$. Finally

$$h_4''(0) = -D_{h_4(0)}Y \cdot h_4'(0) + \left[\frac{d}{dt}\left[\left(\frac{\partial e^{-tY}}{\partial x}\right)_{(t,h_3(t))} \cdot h_3'(t)\right]\right]_{t=0}$$

$$= \left[\frac{d}{dt} \left(\frac{\partial e^{-tY}}{\partial x} \right)_{(t,h_3(t))} \right]_{t=0} \cdot h_3'(0) + \left(\frac{\partial e^{-tY}}{\partial x} \right)_{(0,h_3(0))} \cdot h_3''(0)$$

$$= -D_x Y \cdot Y(x) - 2 D_x X \cdot Y(x) + D_x Y \cdot \left(2X(x) + Y(x) \right)$$

$$= 2 \left(D_x Y \cdot X(x) - D_x X \cdot Y(x) \right) = 2[X,Y](x),$$

which concludes the proof.   □

Remark 1.1 The Lie bracket is bilinear, skew-symmetric and satisfies the Jacobi identity, that is given three smooth vector fields X, Y, Z, we have

$$\big[X,[Y,Z]\big] + \big[Y,[Z,X]\big] + \big[Z,[X,Y]\big] = 0.$$

Remark 1.2 Given a smooth diffeomorphism ϕ from a smooth manifold $\mathcal{U}$ to a smooth manifold $\mathcal{V}$ and X a smooth vector field on $\mathcal{U}$, we recall that the push-forward $\phi_*(X)$ of X is defined by

$$\phi_*(X)(y) := D_{\phi^{-1}(y)} \phi \big(X(\phi^{-1}(y)) \big) \qquad \forall y \in \mathcal{V}.$$

Then if Y ia another smooth vector field on $\mathcal{U}$, we have

$$[\phi_*(X), \phi_*(Y)] = \phi_* ([X,Y]).$$

For any family $\mathcal{F}$ of smooth vector fields on an open set $\mathcal{O} \subset M$, we denote by $\mathrm{Lie}(\mathcal{F})$ the *Lie algebra* of vector fields generated by $\mathcal{F}$. It is the smallest vector subspace S of $\mathcal{X}^\infty(M)$ (the space of smooth vector fields on M) containing $\mathcal{F}$ that also satisfies

$$[X,Y] \in S \qquad \forall X \in \mathcal{F}, \forall Y \in S.$$

It can be constructed as follows: Denote by $\mathrm{Lie}^1(\mathcal{F})$ the space spanned by $\mathcal{F}$ in $\mathcal{X}^\infty(M)$ and define recursively the spaces $\mathrm{Lie}^k(\mathcal{F})$ $(k = 1, 2, \ldots)$ by

$$\mathrm{Lie}^{k+1}(\mathcal{F}) = \mathrm{Span}\left(\mathrm{Lie}^k(\mathcal{F}) \cup \big\{ [X,Y] \,|\, X \in \mathcal{F}, Y \in \mathrm{Lie}^k(\mathcal{F}) \big\} \right) \qquad \forall k \geq 1.$$

This defines an increasing sequence of vector spaces in $\mathcal{X}^\infty(M)$ satisfying

$$\mathrm{Lie}(\mathcal{F}) = \bigcup_{k \geq 1} \mathrm{Lie}^k(\mathcal{F}).$$

In general, $\mathrm{Lie}(\mathcal{F})$ is an infinite-dimensional subspace of $\mathcal{X}^\infty(M)$.

Example 1.5 Let A be a $n \times n$ real matrix, b be a vector in $\mathbb{R}^n$, and X, Y be the smooth vector fields in $\mathbb{R}^n$ defined by

$$X(x) = Ax, \quad Y(x) = b \qquad \forall x \in \mathbb{R}^n.$$

The non-zero Lie brackets of X and Y are always constant vector fields of the form

$$\text{ad}_X^0(Y) := Y = b \quad \text{and} \quad \text{ad}_X^{k+1}(Y) := \left[X, \text{ad}_X^k(Y)\right] = (-1)^{k+1} A^{k+1} b \quad \forall k \geq 0.$$

By the Cayley-Hamilton Theorem, A^n can be expressed as a linear combination of $A^0, \ldots, A^{n-1}$. Therefore, $\text{Lie}(X, Y)$ is the set of vector fields Z in $\mathbb{R}^n$ of the form

$$Z(x) = \lambda A x + \sum_{i=0}^{n-1} \lambda_i A^i b \quad \forall x \in \mathbb{R}^n,$$

with $\lambda, \lambda_0, \ldots, \lambda_{n-1} \in \mathbb{R}$. It is a finite-dimensional Lie algebra.

Example 1.6 Let X, Y be the two smooth vector fields in $\mathbb{R}^2$ (with coordinates $x = (x_1, x_2)$) defined by

$$X(x) = \partial_{x_1}, \quad Y(x) = f(x_1) \partial_{x_2} \quad \forall x \in \mathbb{R}^2,$$

where f is a smooth scalar function. Then, $\text{Lie}(X, Y)$ is the space of smooth vector fields spanned by X and

$$\text{ad}_Y^k(X) = f^{(k)} \partial_{x_2} \quad \text{for } k \geq 0.$$

Thus, $\text{Lie}(X, Y)$ is infinite-dimensional whenever the derivatives of f span an infinite-dimensional space of functions.

For any point $x \in M$, $\text{Lie}(\mathscr{F})(x)$ denotes the set of all tangent vectors $X(x)$ with $X \in \text{Lie}(\mathscr{F})$. It follows that $\text{Lie}(\mathscr{F})(x)$ is always a linear subspace of $T_x M$, hence finite-dimensional.

Example 1.7 Returning to Example 1.6 and denoting by (e_1, e_2) the canonical basis of $\mathbb{R}^2$, we check that

$$\text{Lie}(X, Y)(x) = \text{Span}\left\{e_1, f^{(k)}(x_1) e_2 \mid k = 0, 1, 2, \ldots\right\} \quad \forall x \in \mathbb{R}^2.$$

In particular, $\text{Lie}(X, Y)(x) = \mathbb{R} e_1$ if $f(x)$ and all its derivatives at x vanish and $\text{Lie}(X, Y)(x) = \mathbb{R}^2$ otherwise.

We say that the smooth vector fields $X^1, \ldots, X^m$ satisfy the *Hörmander condition* on some open set $\mathcal{O} \subset M$ if and only if

$$\text{Lie}\left\{X^1, \ldots, X^m\right\}(x) = T_x M \quad \forall x \in \mathcal{O}.$$

A distribution Δ on M is called *totally nonholonomic* on M if for every $x \in M$, there are an open neighborhood $\mathcal{V}_x$ of x in M and a local frame $X_x^1, \ldots, X_x^m$ on $\mathcal{V}_x$ which satisfies the Hörmander condition on $\mathcal{V}_x$. Moreover, we call *degree of nonholonomy* (or simply degree) of Δ at x the smallest integer $r = r(x) \geq 1$ such that

$$\text{Lie}^r\left\{X^1, \ldots, X^m\right\}(x) = T_x M.$$

These definitions are intrinsic, they do not depend upon the choice of the local frame $X_x^1, \ldots, X_x^m$. This is a consequence of the following result:

Proposition 1.6 *Let* $\{X^1, \ldots, X^m\}, \{Y^1, \ldots, Y^m\}$ *be two families of linearly independent smooth vector fields which generate the same distribution on an open set* $\mathcal{O} \subset M$. *Then there holds for any integer* $k \geq 1$,

$$\text{Lie}^k\left\{X^1, \ldots, X^m\right\}(x) = \text{Lie}^k\left\{Y^1, \ldots, Y^m\right\}(x) \quad \forall x \in \mathcal{O}.$$

Proof It is sufficient to show that the left-hand side is included in the right-hand side for any integer $k \geq 2$. Since the $Y^j(x)$ are always linearly independent, there are smooth functions $\alpha_i^j : \mathcal{O} \to \mathbb{R}$ with $i, j = 1, \ldots, m$, such that

$$X^i(x) = \sum_{j=1}^m \alpha_i^j(x) Y^j(x) \quad \forall x \in \mathcal{O}, \quad \forall i = 1, \ldots, m.$$

Then for every $i = 1, \ldots, m$ and every smooth vector field Z, there holds

$$[X^i, Z] = \left[\sum_{j=1}^m \alpha_i^j Y^j, Z\right] = \sum_{j=1}^m \alpha_i^j [Y^j, Z] - \sum_{j=1}^m d\alpha_i^j(Z) Y^j.$$

Since $\text{Span}\left\{X^1(x), \ldots, X^m(x)\right\} \subset \text{Span}\left\{Y^1(x), \ldots, Y^m(x)\right\}$ for any x, this shows that

$$\text{Lie}^2\left\{X^1, \ldots, X^m\right\}(x) \subset \text{Lie}^2\left\{Y^1, \ldots, Y^m\right\}(x) \quad \forall x \in \mathcal{O}.$$

We conclude easily by an inductive argument. □

Remark 1.3 Since for any smooth vector field X, there holds $[X, X] = 0$, a one dimensional distribution cannot be totally nonholonomic.

Example 1.8 The distribution given in Example 1.2 is totally nonholonomic. We check easily that
$$[X, Y] = \partial_z \quad \forall i, j = 1, \ldots, n,$$

which means that Δ has degree 2 everywhere.

Example 1.9 More generally, the distribution given in Example 1.3 is totally nonholonomic of degree 2. We check easily that

$$[X^i, Y^j] = \delta_{ij} \partial_z \qquad \forall i, j = 1, \ldots, n.$$

Example 1.10 The *Martinet distribution* in $\mathbb{R}^3$ (with coordinates (x, y, z)) is the distribution generated by X and Y with

$$X = \partial_x, \quad Y = \partial_y + \frac{x^2}{2} \partial_z.$$

The first Lie bracket of X, Y is given by

$$[X, Y] = x \partial_z.$$

For any $(x, y, z) \in \mathbb{R}^3$ with $x \neq 0$, the three vectors $X(x, y, z), Y(x, y, z)$, $[X, Y](x, y, z)$ are linearly independent. Hence, Δ is a totally nonholonomic distribution of degree 2 on $\mathbb{R}^3 \setminus \{x = 0\}$. Moreover, since $[[X, Y], Y] = \partial_z$, Δ has degree three on the plane $\{x = 0\}$.

Example 1.11 More generally, if X, Y are given by

$$X = \partial_x, \quad Y = \partial_y + x^l \partial_z,$$

with $l \in \mathbb{N}^*$, we check easily that the distribution generated by X and Y is a totally nonholonomic distribution of degree $l + 1$.

Example 1.12 Assume that M has dimension $n = 2p + 1$ and let α be a 1-form on M satisfying

$$\alpha \wedge (d\alpha)^p \neq 0$$

then the distribution given by $\Delta = \mathrm{Ker}(\alpha)$ is totally nonholonomic of degree 2. Such a 1-form is called a *contact form* and the associated distribution is called a *contact distribution*. As a matter of fact, given $\bar{x} \in M$, there is a local set of coordinates $(x_1, \ldots, x_n)$ in an open neighborhood $\mathscr{V}$ of $\bar{x}$ such that α has the form

$$\alpha = \left(\sum_{i=1}^{2p} a_i dx_i \right) + dx_n,$$

where $a_1, \ldots, a_{2p}$ are smooth scalar function on $\mathscr{V}$ such that

$$a_i(\bar{x}) = 0 \qquad \forall i = 1, \ldots, 2p.$$

Hence, the family of smooth vector fields $\bar{X}^1, \ldots, \bar{X}^{2p}$ given by

$$\bar{X}^i = \partial_{x_i} - a_i \partial_{x_n} \qquad \forall i = 1, \ldots, 2p,$$

defines a local frame for $\Delta = \text{Ker}(\alpha)$ in $\bar{\mathscr{V}}$. On the one hand, the $n = 2p + 1$-form $\alpha \wedge (d\alpha)^p$ at $\bar{x}$ reads

$$\left(\alpha \wedge (d\alpha)^p\right)_{\bar{x}}$$

$$= \sum_{\sigma \in \mathscr{P}_{2p}} \left[\prod_{l=1,\dots,p} \left(\frac{\partial a_{j_l}}{\partial x_{i_l}} - \frac{\partial a_{i_l}}{\partial x_{j_l}}\right)\right] dx_n \wedge (dx_{i_1} \wedge dx_{j_1}) \dots \wedge (dx_{i_p} \wedge dx_{j_p}) |_{\bar{x}},$$

(1.3)

where $\mathscr{P}_{2p}$ denotes the set of p-tuples of the form $\sigma = ((i_1, j_1), \dots, (i_p, j_p))$ with $\{i_1, j_1, \dots, i_p, j_p\} = \{1, \dots, 2p\}$ and $i_l < j_l$ for all $l = 1, \dots, p$. On the other hand, we check easily that

$$\left[\bar{X}^i, \bar{X}^j\right](\bar{x}) = \left(\partial_{x_i} a_j - \partial_{x_j} a_i\right) \partial_{x_n}(\bar{x}) \qquad \forall i, j = 1, \dots, 2p.$$

Therefore, if there is $\bar{i} \in \{1, \dots, 2p\}$ such that $[\bar{X}^{\bar{i}}, \bar{X}^j](\bar{x}) = 0$ for any j, then all the products appearing in (1.3) vanish, which implies that $(\alpha \wedge (d\alpha)^p)_{\bar{x}} = 0$, contradiction. We deduce that for every $i \in \{1, \dots, n\}$, there holds

$$\text{Span}\left\{\bar{X}^1(\bar{x}), \dots, \bar{X}^{2p}(\bar{x}), [\bar{X}^i, \bar{X}^1](\bar{x}), \dots, [\bar{X}^i, \bar{X}^{2p}](\bar{x})\right\} = T_{\bar{x}}M. \qquad (1.4)$$

This means that $\Delta = \text{Ker}(\alpha)$ is totally nonholonomic of degree 2.

1.2 Horizontal Paths and End-Point Mappings

Horizontal paths. Let Δ be a distribution of rank $m \leq n$ on M. A continuous path $\gamma : [0, T] \to \mathbb{R}^n$ is said to be an *horizontal* path with respect to Δ if it is absolutely continuous with square integrable derivative (see Appendix A) and satisfies

$$\dot{\gamma}(t) \in \Delta\left(\gamma(t)\right) \quad \text{a.e. } t \in [0, T].$$

For every $x \in M$ and every $T > 0$, we denote by $\Omega_\Delta^{x,T}$ the set of horizontal paths $\gamma : [0, T] \to M$ starting at x. If Δ admits a global frame $X^1, \dots, X^m$, then there is a one-to-one correspondence between $\Omega_\Delta^{x,T}$ and an open subset of $L^2([0, T]; \mathbb{R}^m)$.

Proposition 1.7 *Let $\mathscr{F} = \{X^1, \dots, X^m\}$ be a global frame for Δ. Then for every $x \in M$ and every $T > 0$, there is an open subset $U_{\mathscr{F}}^{x,T}$ of $L^2([0, T]; \mathbb{R}^m)$ such that the mapping*

$$u \in U_{\mathscr{F}}^{x,T} \longmapsto \gamma_u \in \Omega_\Delta^{x,T},$$

(where $\gamma_u : [0, T] \rightarrow M$ is the unique solution to the Cauchy problem

$$\dot{\gamma}_u(t) = \sum_{i=1}^{m} u_i(t) X^i (\gamma_u(t)) \quad a.e. \ t \in [0, T], \qquad \gamma_u(0) = x) \qquad (1.5)$$

is one-to-one.

Proof The set of controls $u \in L^2 ([0, T]; \mathbb{R}^m)$ such that the solution γ_u of (1.5) is well-defined on $[0, T]$ is a non-empty open set. Moreover, by construction, any path γ_u is absolutely continuous with square integrable derivative and almost everywhere tangent to Δ. This proves that the map under study is well-defined on some open set $U_{\mathscr{F}}^{x,T} \subset L^2 ([0, T]; \mathbb{R}^m)$. Let $\gamma \in \Omega_{\Delta, x, T}$ be such that there are $u, v \in L^2 ([0, T]; \mathbb{R}^m)$ with

$$\dot{\gamma}(t) = \sum_{i=1}^{m} u_i(t) X^i (\gamma(t)) = \sum_{i=1}^{m} v_i(t) X^i (\gamma(t)) \qquad a.e. \ t \in [0, T].$$

Since the tangent vectors $X^1 (\gamma(t)), \ldots, X^m (\gamma(t))$ are always linearly independent in $T_{\gamma(t)}M$, we infer that $u(t) = v(t)$ for almost every $t \in [0, T]$, which proves that our map is injective. Furthermore, given $\gamma \in \Omega_{\Delta}^{x,T}$, for almost every $t \in [0, T]$, the path γ is differentiable at t and there is a unique $u(t) \in \mathbb{R}^m$ such that $\dot{\gamma}(t) = \sum_{i=1}^{m} u_i(t) X^i (\gamma(t))$. By construction, the function $u : [0, T] \rightarrow \mathbb{R}^m$ belongs to $L^2 ([0, T]; \mathbb{R}^m)$. $\square$

Remark 1.4 If M is compact, then solutions to (1.5) are defined for any $u \in L^2([0, T]; \mathbb{R}^m)$, which means that $U_{\mathscr{F}}^{x,T} = L^2([0, T]; \mathbb{R}^k)$.

Given a family of smooth vector fields $\mathscr{F} = \{X^1, \ldots, X^k\}$ on M and $x \in M, T > 0$, a function $u \in U_{\mathscr{F}}^{x,T} \subset L^2([0, T]; \mathbb{R}^k)$ is called a *control* and the solution $\gamma_u : [0, T] \rightarrow M$ to the Cauchy problem

$$\dot{\gamma}_u(t) = \sum_{i=1}^{k} u_i(t) X^i (\gamma_u(t)) \quad a.e. \ t \in [0, T], \qquad \gamma_u(0) = x \qquad (1.6)$$

is called the *trajectory* starting at x and associated with the control u. Since any horizontal path can be viewed as a trajectory associated to a control system like (1.6), we restrict in the next paragraph our attention to End-Point mappings associated with finite families of smooth vector fields.

End-Point mappings. Let $\mathscr{F} = \{X^1, \ldots, X^k\}$ be a family of $k \geq 1$ smooth vector fields on M. As before, given x and $T > 0$, there is a maximal open subset $U_{\mathscr{F}}^{x,T} \subset L^2 ([0, T]; \mathbb{R}^k)$ such that for every $u \in U_{\mathscr{F}}^{x,T}$, there is a unique solution to the Cauchy problem (1.6). The *End-Point mapping* associated to $\mathscr{F}$ at x in time $T > 0$ is defined as follows,

$$E_{\mathscr{F}}^{x,T} : U_{\mathscr{F}}^{x,T} \longrightarrow M$$
$$u \longmapsto \gamma_u(T).$$

Given $u \in U_{\mathscr{F}}^{x,T}$, we denote by $X_{\mathscr{F}}^u$ the time-dependent vector field defined by

$$X_{\mathscr{F}}^u(t, x) := \sum_{i=1}^{m} u_i(t) X^i(x) \qquad \text{a.e. } t \in [0, T], \ \forall x \in M.$$

Its flow $\Phi_{\mathscr{F}}^u(t, x)$ is well-defined and smooth on a neighbourhood of x; we denote by $D_x \Phi_{\mathscr{F}}^u(t, x)$ its differential at (t, x) with respect to the x variable. The following result holds. (We refer the reader to Appendix A for reminders in differential equations and to Appendix B for reminders in differential calculus in infinite dimension.)

Proposition 1.8 *The End-Point mapping $E_{\mathscr{F}}^{x,T}$ is of class C^1 on $U_{\mathscr{F}}^{x,T}$ and for every control $u \in U_{\mathscr{F}}^{x,T}$, its differential at u,*

$$D_u E_{\mathscr{F}}^{x,T} : L^2([0, T]; \mathbb{R}^k) \longrightarrow T_{E_{\mathscr{F}}^{x,T}(u)} M$$

is given by

$$D_u E_{\mathscr{F}}^{x,T}(v) = D_x \Phi_{\mathscr{F}}^u(T, x) \cdot \int_0^T \left(D_x \Phi_{\mathscr{F}}^u(t, x) \right)^{-1} \cdot X_{\mathscr{F}}^v \left(t, E_{\mathscr{F}}^{x,t}(u) \right) dt \qquad (1.7)$$

for every $v \in L^2([0, T]; \mathbb{R}^k)$.

Proof Any smooth manifold can be smoothly embedded in an Euclidean space. Then without loss of generality we can assume that M is a smooth submanifold of some $\mathbb{R}^N$ and consequently that the X^i's are the restrictions of smooth vector fields $\tilde{X}^1, \ldots, \tilde{X}^k$ which are defined in an open neighborhood of M in $\mathbb{R}^N$. Given $u \in U_{\mathscr{F}}^{x,T}$ and $v \in L^2([0, T]; \mathbb{R}^k)$ let us look at

$$\lim_{\varepsilon \to 0} \frac{1}{\varepsilon} \left(E_{\mathscr{F}}^{x,T}(u + \varepsilon v) - E_{\mathscr{F}}^{x,T}(u) \right).$$

Using the previous notations, we have

$$\gamma_{u+\varepsilon v}(T) = \int_0^T \sum_{i=1}^{k} (u_i(t) + \varepsilon v_i(t)) \tilde{X}^i \left(\gamma_{u+\varepsilon v}(t) \right) dt, \qquad (1.8)$$

with $\gamma_{u+\varepsilon v}(0) = x$. For every $i = 1, \ldots, k$ and every $t \in [0, T]$, the Taylor expansion of each $\tilde{X}^i$ at $\gamma_u(t)$ gives

$$\tilde{X}^i \left(\gamma_{u+\varepsilon v}(t) \right) = \tilde{X}^i \left(\gamma_u(t) \right) + D_{\gamma_u(t)} \tilde{X}^i \cdot \left(\gamma_{u+\varepsilon v}(t) - \gamma_u(t) \right) + |\gamma_{u+\varepsilon v}(t) - \gamma_u(t)| o(1).$$

Setting $\delta_x(t) := \gamma_{u+\varepsilon v}(t) - \gamma_u(t)$ for any t, we may assume that δ_x has size ε, then (1.8) yields formally

$$\delta_x(T) = \int_0^T \left(\sum_{i=1}^k u_i(t)D_{\gamma_u(t)}\tilde{X}^i \cdot \delta_x(t) + \sum_{i=1}^m \varepsilon v_i(t)\tilde{X}^i\big(\gamma_u(t)\big) \right) dt + o(\varepsilon).$$

This suggests that the linear part in ε of the function $t \in [0, T] \mapsto \delta_x(t)$ should be solution to the Cauchy problem

$$\dot{\xi}(t) = \left[\sum_{i=1}^k u_i(t)D_{\gamma_u(t)}\tilde{X}^i \right]\xi(t) + \left[\sum_{i=1}^k v_i(t)\tilde{X}^i\big(\gamma_u(t)\big) \right] \quad \text{a.e. } t \in [0, T], \qquad (1.9)$$

with $\xi(0) = 0$. Using Gronwall's Lemma (see Appendix A) we check easily that for every $v \in L^2\big([0, T]; \mathbb{R}^k\big)$, the quantity

$$\frac{1}{\varepsilon}\left(E_{\mathscr{F}}^{x,T}(u + \varepsilon v) - E_{\mathscr{F}}^{x,T}(u) - \varepsilon\xi(T) \right)$$

tends to zero as ε tends to zero. For almost every $t \in [0, T]$, denote by $A_u(t)$ the matrix in $M_N(\mathbb{R})$ representing the linear operator $\sum_{i=1}^k u_i(t)D_{\gamma_u(t)}\tilde{X}^i$ in the canonical basis of $\mathbb{R}^N$ and for every $t \in [0, T]$, denote by $B_u(t)$ the matrix in $M_{N,k}(\mathbb{R})$ whose columns are the $\tilde{X}^i(\gamma_u(t))$'s. Denote by $S_u : [0, T] \to M_N(\mathbb{R})$ the solution to the Cauchy problem

$$\dot{S}_u(t) = A_u(t)S_u(t) \quad \text{a.e. } t \in [0, T], \quad S_u(0) = I_n.$$

Note that $S_u(t)$ is exactly the Jacobian of the flow $\Phi_{\tilde{\mathscr{F}}}^u$ (with $\tilde{\mathscr{F}} = \{\tilde{X}^1, \dots, \tilde{X}^k\}$) at $(t, \gamma_u(t))$ with respect to the x variable. The solution of (1.9) at time T is given by (see Appendix A)

$$\xi(T) = D_u E_{\mathscr{F}}^{x,T}(v) = S_u(T)\int_0^T S_u(t)^{-1}B_u(t)v(t)dt.$$

Thus $E_{\mathscr{F}}^{x,T}$ is differentiable at u and (1.7) is satisfied. Using Gronwall's lemma again, we leave the reader to check that $D_u E_{\mathscr{F}}^{x,T}$ depends continuously on u on $U_{\mathscr{F}}^{x,T}$. $\quad\square$

Remark 1.5 If $M = \mathbb{R}^n$, the derivative of $E_{\mathscr{F}}^{x,T}$ at u is given by

$$D_u E_{\mathscr{F}}^{x,T}(v) = S(T)\int_0^T S(t)^{-1}B(t)v(t)dt,$$

where $S : [0, T] \to M_n(\mathbb{R})$ is the solution to the Cauchy problem

$$\dot{S}(t) = A(t)S(t) \quad \text{a.e. } t \in [0, T], \quad S(0) = I_n \qquad (1.10)$$

and where the matrices $A(t) \in M_n(\mathbb{R})$, $B(t) \in M_{n,k}(\mathbb{R})$ are defined by

$$A(t) := \sum_{i=1}^{k} u_i(t) J_{X^i}(\gamma_u(t)) \qquad \text{a.e. } t \in [0, T] \tag{1.11}$$

$(\gamma_u(t) = E_{\mathscr{F}}^{x,t}(u)$ and J_{X^i} denotes the Jacobian matrix of X^i at $\gamma_u(t))$ and

$$B(t) := \left(X^1(\gamma_u(t)), \dots, X^k(\gamma_u(t))\right) \qquad \forall t \in [0, T]. \tag{1.12}$$

Properties of End-Point mappings. Given $u \in U_{\mathscr{F}}^{x,T}$, we set

$$\text{Im}_{\mathscr{F}}^{x,T}(u) := D_u E_{\mathscr{F}}^{x,T}\left(L^2([0, T]; \mathbb{R}^k)\right).$$

Defining $y = E_{\mathscr{F}}^{x,T}(u)$, we observe that $\text{Im}_{\mathscr{F}}^{x,T}(u)$ is a vector space contained in $T_y M$, hence of dimension $\leq n$. We call *rank* of $u \in U_{\mathscr{F}}^{x,T}$ with respect to $E_{\mathscr{F}}^{x,T}$, denoted by $\text{rank}_{\mathscr{F}}^{x,T}(u)$, the dimension of $\text{Im}_{\mathscr{F}}^{x,T}(u)$.

Given $u \in L^2([0, T]; \mathbb{R}^k)$ and $\lambda > 0$, we define the controls $u_\lambda \in L^2([0; \lambda^{-1}T]; \mathbb{R}^k)$ and $\check{u} \in L^2([0, T]; \mathbb{R}^k)$ by

$$u_\lambda(t) := \lambda u(\lambda t) \qquad \text{a.e. } t \in [0, \lambda^{-1}T],$$

$$\text{and} \quad \check{u}(t) := -u(T - t) \qquad \text{a.e. } t \in [0, T].$$

Moreover, if in addition $u' \in L^2([0, T]; \mathbb{R}^k)$ we define $u * u'$, the *concatenation* of u and u', in $L^2[0, T + T']; \mathbb{R}^k)$ by

$$u * u'(t) = \begin{cases} u(t) & \text{if } 0 \leq t \leq T; \\ u'(t - T) & \text{if } T < t \leq T + T' \end{cases} \qquad \text{a.e. } t \in [0, T + T'].$$

Proposition 1.9 *Let $u \in U_{\mathscr{F}}^{x,T}$, $u \in U_{\mathscr{F}}^{x,T'}$ and $\lambda > 0$ be fixed, then we have (we set $y := E_{\mathscr{F}}^{x,T}(u))$:*

(i) $u_\lambda \in U_{\mathscr{F}}^{x,\lambda^{-1}T}$ *and* $\text{rank}_{\mathscr{F}}^{x,T}(u) = \text{rank}_{\mathscr{F}}^{x,\lambda^{-1}T}(u_\lambda)$.

(ii) $\check{u}$ *belongs to* $U_{\mathscr{F}}^{y,T}$ *and* $\text{rank}_{\mathscr{F}}^{x,T}(u) = \text{rank}_{\mathscr{F}}^{y,T}(\check{u})$.

(iii) $u * u' \in U_{\mathscr{F}}^{x,T+T'}$ *and* $\text{rank}_{\mathscr{F}}^{x,T+T'}(u * u') \geq \max\left\{\text{rank}_{\mathscr{F}}^{x,T}(u), \text{rank}_{\mathscr{F}}^{y,T'}(u')\right\}$.

Proof To prove (i), we note that if $\gamma_u : [0, T] \to M$ is a solution of (1.6), then the path $\gamma_{u,\lambda} : [0, \lambda^{-1}T] \to M$ defined by $\gamma_{u,\lambda}(t) = \gamma_u(\lambda t)$ for any $t \in [0, \lambda^{-1}T]$, satisfies for a.e. $t \in [0, \lambda^{-1}T]$,

$$\dot{\gamma}_{u,\lambda}(t) = \lambda \dot{\gamma}_u(\lambda t) = \sum_{i=1}^{k} \lambda u_i(\lambda t) X^i \left(\gamma_u(\lambda t) \right) = \sum_{i=1}^{k} u_\lambda(t) X^i \left(\gamma_{u,\lambda}(t) \right).$$

We infer that u_λ belongs to $U_{\mathscr{F}}^{x,\lambda^{-1}T}$ and satisfies

$$D_{u_\lambda} E_{\mathscr{F}}^{x,\lambda^{-1}T}(v_\lambda) = D_u E_{\mathscr{F}}^{x,T}(v) \qquad \forall v \in L^2\left([0,T]; \mathbb{R}^k\right).$$

We conclude easily. To prove (ii), we note that

$$E_{\mathscr{F}}^{E_{\mathscr{F}}^{x,T}(u),T} \left(\check{u} \right) = x \qquad \forall u \in U_{\mathscr{F}}^{x,T}.$$

The mapping $(z, v) \mapsto E_{\mathscr{F}}^{z,T}(v)$ is C^1 and its derivative with respect to the z variable at $\left(y = E_{\mathscr{F}}^{x,T}(u), \check{u} \right)$ is given by $D_x \Phi_{\mathscr{F}}^u(T, x)^{-1}$. Derivating the above equality at u yields

$$(D_x \Phi_{\mathscr{F}})^u (T, x)^{-1} \cdot D_u E_{\mathscr{F}}^{x,T}(v) + D_{\check{u}} E_{\mathscr{F}}^{y,T} (\check{v}) = 0 \qquad \forall v \in L^2\left([0,T]; \mathbb{R}^k\right).$$

The result follows easily. We now observe that

$$E_{\mathscr{F}}^{x,T+T'}(u * u') = E_{\mathscr{F}}^{E_{\mathscr{F}}^{x,T}(u),T'}(u'),$$

for any $u \in U_{\mathscr{F}}^{x,T}$ and $u' \in U_{\mathscr{F}}^{y,T'}$ with $y = E_{\mathscr{F}}^{x,T}(u)$. Again, derivating yields

$$D_{u*u'} E_{\mathscr{F}}^{x,T+T'}(v * v') = D_x \Phi_{\mathscr{F}}^{u'}(T', y) \cdot D_u E_{\mathscr{F}}^{x,T}(v) + D_{u'} E_{\mathscr{F}}^{y,T'}(v'), \qquad (1.13)$$

for any $v \in L^2([0,T]; \mathbb{R}^k)$ and $v' \in L^2([0,T']; \mathbb{R}^k)$. We conclude easily. $\square$

As the next example shows, the inequality in Proposition 1.9 may be strict.

Example 1.13 Let $\mathscr{F} = \{X^1, X^2\}$ be the family of smooth vectors fields on $\mathbb{R}^4$ (with coordinates $x = (x_1, x_2, x_3, x_4)$ and canonical basis (e_1, e_2, e_3, e_4)) defined by

$$X^1 = \partial_{x_1} \quad \text{and} \quad X^2 = \partial_{x_2} + x_1^2 \partial_{x_3} + x_1 x_2 \partial_{x_4}.$$

Set $x = (-1, 0, 0, 0)$, $y = (0, 0, 0, 0)$, and define the controls $u, u' \in L^2([0, 1]; \mathbb{R}^2)$ by

$$u(t) = (1, 0) \quad \text{and} \quad u'(t) = (0, 1) \qquad \forall t \in [0, 1].$$

The control u has rank 3 with respect to $E_{\mathscr{F}}^{x,1}$. As a matter of fact, the trajectory $\gamma_u : [0, 1] \to \mathbb{R}^4$ starting at x and associated with u equals $\gamma_u(t) = (-1 + t, 0, 0, 0)$ for any $t \in [0, 1]$ and using the representation formula given in Remark 1.5, we have

$$D_u E^{x,1}_{\mathscr{F}}(v) = \int_0^T B(t)v(t)dt \quad \forall v \in L^2\big([0,1]; \mathbb{R}^2\big),$$

where $B(t) = \big(X^1(\gamma_u(t)), X^2(\gamma_u(t))\big)$ for any $t \in [0,1]$. Then,

$$\text{Im}^{x,1}_{\mathscr{F}}(u) = \text{Span}\left\{ \int_0^1 v_1(t)dt\, e_1 \mid v_1 \in L^2([0,1]; \mathbb{R}) \right\}$$

$$+ \text{Span}\left\{ \int_0^1 v_2(t)dt\, e_2 + \int_0^1 (1-t)^2 v_2(t)dt\, e_3 \mid v_2 \in L^2([0,1]; \mathbb{R}) \right\}$$

$$= \text{Span}\{e_1, e_2, e_3\}.$$

The trajectory $\gamma_{u'} : [0,1] \to \mathbb{R}^4$ starting at y and associated with u' equals $\gamma_{u'}(t) = (0,t,0,0)$ for any $t \in [0,1]$, and there holds

$$D_{u'} E^{y,1}_{\mathscr{F}}(v) = S(T) \int_0^T S(t)^{-1} B(t)v(t)dt,$$

where

$$S(t) = \begin{pmatrix} 1 & 0 & 0 & 0 \\ 0 & 1 & 0 & 0 \\ 0 & 0 & 1 & 0 \\ \frac{t^2}{2} & 0 & 0 & 1 \end{pmatrix} \quad \text{and} \quad B(t) = \begin{pmatrix} 1 & 0 \\ 0 & 1 \\ 0 & 0 \\ 0 & 0 \end{pmatrix} \quad \forall t \in [0,1].$$

We infer that

$$\text{Im}^{y,1}_{\mathscr{F}}(u') = \text{Span}\left\{ \int_0^1 v_2(t)dt\, e_2 \mid v_2 \in L^2([0,1]; \mathbb{R}) \right\}$$

$$+ \text{Span}\left\{ \int_0^1 v_1(t)dt\, e_1 + \int_0^1 \left(1 - \frac{t^2}{2}\right) v_1(t)dt\, e_4 \mid v_1 \in L^2([0,1]; \mathbb{R}) \right\}$$

$$= \text{Span}\{e_1, e_2, e_4\}.$$

Finally, we note that

$$D_x \Phi^{u'}_{\mathscr{F}}(1, y)(e_3) = e_3.$$

Therefore, by (1.13), this implies $\text{Im}^{x,2}_{\mathscr{F}}(u * u') = \mathbb{R}^4$, which means that the rank $\text{rank}^{x,2}_{\mathscr{F}}(u * u') = 4$ is strictly larger than the maximum of $\text{rank}^{x,1}_{\mathscr{F}}(u)$ and $\text{rank}^{y,1}_{\mathscr{F}}(u')$ which is equal to 3.

The following proposition implies that the rank of a control is always larger or equal than the dimension of the family $\{X^1, \ldots, X^k\}$ at the end-point.

Proposition 1.10 *We have for every $u \in U_{\mathscr{F}}^{x,T}$,*

$$X^i\left(E_{\mathscr{F}}^{x,T}(u)\right) \in Im_{\mathscr{F}}^{x,T}(u) \quad \forall i = 1, \ldots, k.$$

Proof Let us first assume that we work in $\mathbb{R}^n$. In this case (see Remark 1.5), the derivative of $E_{\mathscr{F}}^{x,T}$ at u is given by

$$D_u E_{\mathscr{F}}^{x,T}(v) = S(T) \int_0^T S(t)^{-1} B(t) v(t) dt \quad \forall v \in L^2\left([0, T]; \mathbb{R}^k\right),$$

where $S(\cdot)$ is the solution to the Cauchy problem (1.10) and where the matrices $A(t) \in M_n(\mathbb{R}), B(t) \in M_{n,k}(\mathbb{R})$ are defined respectively by (1.11) and (1.12). Fix $i \in \{1, \ldots, k\}$ and denote by e_i the i-th vector of the canonical basis in $\mathbb{R}^k$. Define, for every $\varepsilon \in (0, T)$, the control $v_\varepsilon \in L^2\left([0, T]; \mathbb{R}^k\right)$ by

$$v_\varepsilon(t) = \begin{cases} 0 & \text{if } 0 \le t \le T - \varepsilon; \\ (1/\varepsilon)e_i & \text{if } T - \varepsilon < t \le T. \end{cases}$$

We have

$$\left| D_u E_{\mathscr{F}}^{x,T}(v_\varepsilon) - X^i(\gamma_u(T)) \right|$$

$$= \left| (1/\varepsilon)S(T) \int_{T-\varepsilon}^T S(t)^{-1} X^i(\gamma_u(t)) \, dt - (1/\varepsilon)S(T) \int_{T-\varepsilon}^T S(T)^{-1} X^i(\gamma_u(T)) \, dt \right|$$

$$\le (1/\varepsilon) |S(T)| \int_{T-\varepsilon}^T \left| S(t)^{-1} X^i(\gamma_u(t)) - S(T)^{-1} X^i(\gamma_u(T)) \right| dt$$

$$\le (1/\varepsilon) |S(T)| \int_{T-\varepsilon}^T \left\| S(t)^{-1} \right\| \left| X^i(\gamma_u(t)) - X^i(\gamma_u(T)) \right| dt$$

$$+ (1/\varepsilon) |S(T)| \int_{T-\varepsilon}^T \left\| S(t)^{-1} - S(T)^{-1} \right\| \left| X^i(\gamma_u(T)) \right| dt.$$

Both mappings $t \mapsto X^i(x_u(t))$ and $t \mapsto S(t)^{-1}$ are continuous at $t = T$. Therefore, there holds

$$\lim_{\varepsilon \downarrow 0} D_u E_{\mathscr{F}}^{x,T}(v_\varepsilon) = X^i(\gamma_u(T)).$$

Since $Im_{\mathscr{F}}^{x,T}(u) = D_u E_{\mathscr{F}}^{x,T}\left(L^2([0, T]; \mathbb{R}^k)\right)$ is a closed subset of $\mathbb{R}^k$, we infer that $X^i(\gamma_u(T))$ belongs to $Im_{\mathscr{F}}^{x,T}(u)$.

If we are now in M, then there exists a local chart around x and $\bar{t} \in (0, T)$ such that $\gamma_u(\bar{t}) \in \mathcal{O}$. Set $T' := T - \bar{t}$ and define $u^1 \in L^2([0, \bar{t}]; \mathbb{R}^k)$ and $u^2 : L^2([0, T']; \mathbb{R}^k)$ by

$$u^1(t) = u(t) \quad \forall t \in [0, \bar{t}] \quad \text{and} \quad u^2(t) = u(t + \bar{t}) \quad \forall t \in [0, T'].$$

We conclude easily by the above proof in $\mathbb{R}^n$ together with (1.13). $\square$

1.3 Regular and Singular Horizontal Paths

Regular and singular controls. Let $\mathcal{F} = \{X^1, \ldots, X^k\}$ be a family of $k \geq 1$ smooth vector fields on M. Given $x \in M$ and $T > 0$, we say that the control $u \in U_{\mathcal{F}}^{x,T}$ is *regular* with respect to x and $\mathcal{F}$ if $\mathrm{rank}_{\mathcal{F}}^{x,T}(u) = n$ (recall that M has dimension n). Otherwise, we shall say that u is *singular*. In other terms, u is singular if and only if it is a critical point of the End-Point mapping $E_{\mathcal{F}}^{x,T}$, that is if $E_{\mathcal{F}}^{x,T}$ is not a submersion at u.

Remark 1.6 Proposition 1.10 shows that if $\mathcal{F} = \{X^1, \ldots, X^k\}$ is a family of smooth vector fields on M such that

$$\mathrm{Span}\left\{X^1(x), \ldots, X^k(x)\right\} = T_x M \quad \forall x \in M,$$

then every non-trivial admissible control (that is u in some $U_{\mathcal{F}}^{x,T}$ with $u \neq 0, T > 0$) is regular.

Propositions 1.9 shows that a given control $u \in U_{\mathcal{F}}^{x,T}$ is singular with respect to x and $\mathcal{F}$ if and only if any control of the form $u_\lambda \in U_{\mathcal{F}}^{x,\lambda^{-1}T}$ (with $\lambda \neq 0$) is singular with respect to x and $\mathcal{F}$ and if and only if $\check{u} \in U_{\mathcal{F}}^{y,T}$ (with $y = E_{\mathcal{F}}^{x,T}$) is singular with respect to y and $\mathcal{F}$. It also shows that if the concatenation of several controls is singular then each of them is singular.

Define k Hamiltonians $h^1, \ldots, h^k : T^*M \to \mathbb{R}$ by $h^i := h_{X^i}$ for any $i = 1, \ldots, k$, that is

$$h^i(\psi) = p \cdot X^i(x) \quad \forall \psi = (x, p) \in T^*M, \quad \forall i = 1, \ldots, k.$$

For every $i = 1, \ldots, m$, $\overrightarrow{h}_i$ denotes the Hamiltonian vector field on T^*M associated to h_i, which in local coordinates on T^*M reads

$$\overrightarrow{h}_i(x, p) = \left(\frac{\partial h_i}{\partial p}(x, p), -\frac{\partial h_i}{\partial x}(x, p) \right).$$

Singular controls can be characterized as follows.

Proposition 1.11 *The control $u \in U_{\mathscr{F}}^{x,T}$ is singular with respect to x and $\mathscr{F}$ if and only if there exists an absolutely continuous arc $\psi : [0, T] \to T^*M$ that never intersects the zero section of T^*M, such that*

$$\dot{\psi}(t) = \sum_{i=1}^{k} u_i(t) \overrightarrow{h}^{\,i}(\psi(t)) \quad a.e.\ t \in [0, T] \qquad (1.14)$$

and

$$h^i(\psi(t)) = 0, \quad \forall t \in [0, T] \quad \forall i = 1, \ldots, k. \qquad (1.15)$$

We say that ψ is an abnormal extremal lift of $\gamma_u : [0, T] \to M$ (defined by (1.6)).

Proof Let us first assume that we work in $\mathbb{R}^n$. If $D_u E_{\mathscr{F}}^{x,T} : L^2([0, T]; \mathbb{R}^k) \to \mathbb{R}^n$ is not surjective, then there exists $p \in (\mathbb{R}^n)^* \setminus \{0\}$ such that

$$p \cdot D_u E_{\mathscr{F}}^{x,T}(v) = 0 \quad \forall v \in L^2([0, T]; \mathbb{R}^k).$$

Remembering Remark 1.5, the above identity can be written as

$$\int_0^T pS(T)S(t)^{-1}B(t)v(t)dt = 0 \quad \forall v \in L^2([0, T]; \mathbb{R}^k).$$

Taking $v \in L^2([0, T]; \mathbb{R}^k)$ defined as

$$v(t) = \left(pS(t)S(t)^{-1}B(t) \right)^* \quad \forall t \in [0, T],$$

we deduce that

$$\int_0^T \left| \left(pS(T)S(t)^{-1}B(t) \right)^* \right|^2 ds = 0,$$

which implies that $pS(T)S(t)^{-1}B(t) = 0$ for any $t \in [0, T]$ (note that the function $t \mapsto pS(T)S(t)^{-1}B(t)$ is continuous). Let us now define, for each $t \in [0, T]$,

$$p(t) := pS(T)S(t)^{-1}.$$

By construction, $p : [0, T] \to (\mathbb{R}^n)^*$ is an absolutely continuous arc. Since $p \neq 0$ and $S(t)$ is invertible for all $t \in [0, T]$, $p(t)$ does not vanish on $[0, T]$. Moreover, recalling that, by definition of S,

$$\frac{d}{dt}S(t)^{-1} = -S(t)^{-1}A(t) \quad a.e.\ t \in [0, T],$$

we conclude that p satisfies the following properties:

$$\dot{p}(t) = -p(t)A(t) \quad \text{a.e. } t \in [0, T]$$

and

$$p(t)B(t) = 0 \quad \forall t \in [0, T]$$

which shows that (1.14)–(1.15) are satisfied with $\psi(t) = (\gamma_u(t), p(t))$ for any $t \in [0, T)$. By the way, we note that by construction, we have for every $t \in (0, T]$,

$$p(t) \cdot D_{u_t} E_{\mathscr{F}}^{x,t}(v) = 0 \quad \forall v \in L^2\big([0, t]; \mathbb{R}^k\big), \tag{1.16}$$

where u_t denotes the restriction of u to $[0, t]$.

Conversely, let us assume that there exists an absolutely continuous arc $p : [0, T] \to (\mathbb{R}^n)^* \setminus \{0\}$ such that (1.14) and (1.15) are satisfied with $\psi = (\gamma_u, p)$. This means that

$$-\dot{p}(t) = p(t)A(t) \quad \text{a.e. } t \in [0, T]$$

and

$$p(t)^* B(t) = 0 \quad \forall t \in [0, T].$$

Setting $p := p(T) \neq 0$, we have, for any $t \in [0, T]$,

$$p(t) = pS(T)S(t)^{-1}.$$

Hence, we obtain

$$pS(T)S(t)^{-1}B(t) = 0 \quad \forall t \in [0, T],$$

which in turn implies

$$p \cdot D_u E_{\mathscr{F}}^{x,T}(v) = 0, \quad \forall v \in L^2([0, T]; \mathbb{R}^k).$$

This concludes the proof. Again, the above proof shows indeed that (1.16) holds for any $t \in (0, T]$.

Assume now that we work on M. We can cut the path $\gamma_u : [0, T] \to M$ associated with u and u itself into a finite number of pieces $\gamma^1, \ldots, \gamma^l$ and $u^1, \ldots, u^l$ such that each control u^l is singular and each path γ^l is valued in a chart of M. Then we can apply the previous arguments on each chart and thanks to (1.16) obtain a non-vanishing absolutely continuous arc ψ satisfying (1.14)–(1.15) on $[0, T]$. $\quad \square$

Remark 1.7 We keep in mind that if $\psi : [0, T] \to T^*M$ is an absolutely continuous arc satisfying (1.14)–(1.15), then

Fig. 1.2 The concatenation $\gamma^1 * \gamma^2 * \gamma^3$ of three paths

$$p(t) \cdot D_{u_t} E_{\mathscr{F}}^{x,t}(v) = 0 \quad \forall v \in L^2([0, t]; \mathbb{R}^k),$$

where $\psi(t) = (\gamma_u(t), p(t))$ and u_t denotes the restriction of u to $[0, t]$. (In the sequel, $\psi \cdot v$ or $p \cdot v$ with $\psi = (x, p)$ in local coordinates denotes the evaluation of the form ψ at $v \in T_x M$.)

Remark 1.8 In local coordinates, Proposition 1.11 means that there exists an absolutely continuous arc $p : [0, T] \to (\mathbb{R}^n)^* \setminus \{0\}$ satisfying

$$\dot{p}(t) = -\sum_{i=1}^{k} u_i(t) \, p(t) \cdot D_{\gamma_u(t)} X^i \quad \text{a.e. } t \in [0, T] \tag{1.17}$$

and

$$p(t) \cdot X^i(\gamma_u(t)) = 0 \quad \forall t \in [0, T], \, \forall i = 1, \ldots k. \tag{1.18}$$

Regular and singular paths. Regular Let Δ be a distribution of rank $m \le n$ on M. As seen before, it can be represented by a generating family $\mathscr{F} = \{X^1, \ldots, X^k\}$ of smooth vector fields (see Proposition 1.3). Given a point $x \in M$, a time $T > 0$, and an horizontal path $\gamma \in \Omega_{\Delta}^{x,T}$, we set

$$\operatorname{Im}_{\Delta}(\gamma) := D_u E_{\mathscr{F}}^{x,T}\left(L^2([0, T]; \mathbb{R}^k)\right) \subset T_{E_{\mathscr{F}}^{x,T}(u)} M,$$

where $u \in U_{\mathscr{F}}^{x,T}$ is any control such that $\gamma = \gamma_u$ is solution to the Cauchy problem (1.6). We call *rank* of $\gamma \in \Omega_{\Delta}^{x,T}$, denoted by $\operatorname{rank}_{\Delta}(\gamma)$, the dimension of $\operatorname{Im}_{\Delta}(u)$. We shall say that γ is *singular* (with respect to Δ) if $\operatorname{rank}_{\Delta}(\gamma) < n$ and *regular* otherwise.

Remark 1.9 By Remark 1.6, if Δ has rank $m = n$ then any non-trivial horizontal path is regular.

Proposition 1.9 does apply to horizontal paths. The rank of an horizontal path depends only on the curve drawn by the path in M, it does not depend upon its parametrization. Moreover if an horizontal path which is the concatenation of several paths (the concatenation of paths is defined in the same way as the concatenation of controls, see Fig. 1.2) is singular then each piece is necessarily singular.

Example 1.14 Returning to Examples 1.2 and 1.8, we consider in $\mathbb{R}^3$ with coordinates $x = (x_1, x_2, x_3)$, the totally nonholonomic rank two distribution Δ generated by

$$X^1 = \partial_{x_1} - \frac{x_2}{2}\partial_{x_3} \quad \text{and} \quad X^2 = \partial_{x_2} + \frac{x_1}{2}\partial_{x_3}.$$

We claim that the singular horizontal paths are the constant curves or equivalently that the only singular control with respect to $\mathscr{F} = \{X^1, X^2\}$ is the control $u \equiv 0$. Let us prove this claim. Let $x \in \mathbb{R}^3$, $T > 0$ be fixed and $u \in U_{\mathscr{F}}^{x,T}$ be a singular control. Denote by $x : [0, T] \to \mathbb{R}^3$ the solution to the Cauchy problem

$$\dot{x}(t) = u_1(t)X^1(x(t)) + u_2(t)X^2(x(t)) \quad \text{a.e. } t \in [0, T], \quad x(0) = x. \quad (1.19)$$

From Proposition 1.11, there exists an absolutely continuous arc $p : [0, T] \to \left(\mathbb{R}^3\right)^* \setminus \{0\}$ such that

$$\dot{p}(t) = -u_1(t)p(t) \cdot D_{x(t)}X^1 - u_2(t)p(t) \cdot D_{x(t)}X^2 \qquad (1.20)$$

for a.e. $t \in [0, T]$ and

$$p(t) \cdot X^1(x(t)) = p(t) \cdot X^2(x(t)) = 0 \qquad \forall t \in [0, T]. \qquad (1.21)$$

Taking the derivatives in (1.21) gives

$$\dot{p}(t) \cdot X^i(x(t)) + p(t) \cdot D_{x(t)}X^i\big(\dot{x}(t)\big) = 0 \qquad \text{a.e. } t \in [0, T], \, \forall i = 1, 2$$

which implies, by (1.19)–(1.20),

$$u_1(t)p(t) \cdot [X^1, X^i](x(t)) + u_2(t)p(t) \cdot [X^2, X^i](x(t)) = 0 \qquad \text{a.e. } t \in [0, T].$$

Taking $i = 1$ and $i = 2$, we obtain that for almost every $t \in [0, T]$,

$$u_1(t)p(t) \cdot [X^1, X^2](x(t)) = u_2(t)p(t) \cdot [X^1, X^2](x(t)) = 0.$$

Since $[X^1, X^2] = -\frac{\partial}{\partial x_3}$ and (1.21) is satisfied with $p(t) \neq 0$, we deduce that $u \equiv 0$.

Example 1.15 The property of the previous example is satisfied by much more general distributions. A distribution Δ on M is called *fat* if, for every $x \in M$ and every section X of Δ with $X(x) \neq 0$, there holds

$$T_x M = \Delta(x) + [X, \Delta](x), \qquad (1.22)$$

where

$$[X, \Delta](x) := \Big\{[X, Z](x) \,|\, Z \text{ section of } \Delta\Big\}.$$

The condition above being very restrictive, there are very few fat distributions. Fat distributions on three-dimensional manifolds are the rank-two distributions Δ satisfying

$$T_x M = \mathrm{Span}\left\{X^1(x), X^2(x), [X^1, X^2](x)\right\} \qquad \forall x \in \mathscr{V},$$

where (X^1, X^2) is a local frame for Δ in $\mathscr{V}$. Another example of co-rank one fat distributions in odd dimension is given by contact distributions which were introduced in Example 1.12. In this case property (1.22) is an easy consequence of (1.4). Let us now prove that fat distributions do not admit non-trivial singular horizontal paths. By the property of singular concatenated horizontal paths, we just need to show that non-constant short horizontal paths cannot be singular. Taking a local chart if necessary we can work in $\mathbb{R}^n$ and assume that Δ has a local frame $X^1, \ldots, X^m$. Let $x \in \mathbb{R}^n, T > 0$ be fixed and $u \in U_{\mathscr{F}}^{x,T}$ be a singular control. By Remark 1.8, there exists an absolutely continuous arc $p : [0, T] \to (\mathbb{R}^n)^* \setminus \{0\}$ satisfying (1.17) and (1.18). For almost every fixed $t \in [0, T]$ and every $i = 1, \ldots, m$, derivating (1.18) yields

$$\sum_{j=1}^m u_j(t) \, p(t) \cdot \left[X^j, X^i\right] (\gamma_u(t)) = p(t) \cdot \left[\sum_{j=1}^m u_j(t) X^j, X^i\right] (\gamma_u(t)) = 0.$$

Setting the autonomous vector field $X(\cdot) := \sum_{j=1}^m u_j(t) X^j(\cdot)$, we deduce that $p(t)$ annihilates all the $X^i(\gamma_u(t))$'s and all the $[X, X^i](\gamma_u(t))$'s. This contradicts (1.22).

Example 1.16 Returning to Example 1.10 (Martinet distribution), we consider in $\mathbb{R}^3$ with coordinates $x = (x_1, x_2, x_3)$, the totally nonholonomic rank two distribution Δ generated by

$$X^1 = \partial_{x_1} \quad \text{and} \quad X^2 = \partial_{x_2} + \frac{x_1^2}{2} \partial_{x_3}.$$

We claim that the singular horizontal curves are exactly the "traces of the distribution" on the so-called *Martinet surface* (see Fig. 1.3)

$$\Sigma_\Delta := \left\{x \in \mathbb{R}^3 \mid x_1 = 0\right\},$$

which in other terms means that the singular horizontal paths are either constant curves or are contained in a line l_z of the form

$$l_z = \left\{x = (x_1, x_2, x_3) \in \mathbb{R}^3 \mid x_1 = 0 \text{ and } x_3 = z\right\}$$

for some $z \in \mathbb{R}$.

Let us prove this claim. Let $x \in \mathbb{R}^3, T > 0$ be fixed and $u \in U_{\mathscr{F}}^{x,T}$ be a non-trivial singular control. Denote by $x : [0, T] \to \mathbb{R}^3$ the solution to the Cauchy problem

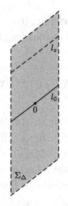

Fig. 1.3 The singular *horizontal curves* are the traces of the distribution on the Martinet surface

$$\dot{x}(t) = u_1(t)X^1(x(t)) + u_2(t)X^2(x(t)) \quad \text{a.e. } t \in [0, T], \quad x(0) = x.$$

As in the previous example, from Proposition 1.11, there exists an absolutely continuous arc $p : [0, T] \to \left(\mathbb{R}^3\right)^* \setminus \{0\}$ such that

$$\dot{p}(t) = -u_1(t)p(t) \cdot D_{x(t)}X^1 - u_2(t)p(t) \cdot D_{x(t)}X^2 \qquad (1.23)$$

for a.e. $t \in [0, T]$ and

$$p(t) \cdot X^1(x(t)) = p(t) \cdot X^2(x(t)) = 0 \qquad \forall t \in [0, T]. \qquad (1.24)$$

We deduce that

$$|u(t)|^2 \left(p(t) \cdot [X^1, X^2](x(t)) \right)^2 = 0 \qquad \text{a.e. } t \in [0, T].$$

Since the three vectors $X^1(x), X^2(x), [X^1, X^2](x)$ span $\mathbb{R}^3$ for every x with $x_1 \neq 0$, this shows that $x_1(t) = 0$ for all $t \in [0, T]$, which in turn implies that $u_1 \equiv 0$. We deduce that x has the form

$$x(t) = \left(0, x_2(0) + \int_0^t u_2(s)ds, 0, x_3(0) \right),$$

which shows that it is contained in $l_{x_3(0)}$. Conversely, if an horizontal path $x \in \Omega_\Delta^{x,T}$ has the form

$$x(t) = (0, x_2(t), z) \qquad \forall t \in [0, T]$$

with $z \in \mathbb{R}$, then any absolutely continuous arc $p : [0, T] \to \mathbb{R}^3 \setminus \{0\}$ of the form

$$p(t) = (0, 0, p_3) \qquad \forall t \in [0, T]$$

with $p_3 \neq 0$ satisfies (1.23) and (1.24). This shows that any horizontal path which is contained in a line l_z for some $z \in \mathbb{R}$ is singular.

Example 1.17 More generally, consider a totally nonholonomic distribution Δ of rank two in a manifold M of dimension three. We define the *Martinet surface* of Δ as the set defined by

$$\Sigma_\Delta := \Big\{ x \in M \mid \Delta(x) + [\Delta, \Delta](x) \neq T_x M \Big\},$$

where

$$\big[\Delta, \Delta\big](x) := \Big\{ [X, Y](x) \mid X, Y \text{ sections of } \Delta \Big\}.$$

In other terms, a point $x \in M$ belongs to Σ_Δ if and only if Δ is not a contact distribution at x, that is if for any (or for only one) local frame $\{X^1, X^2\}$ in a neighborhood of x the three vectors $X^1(x), X^2(x), [X^1, X^2](x)$ do not span $T_x M$. The singular paths with respect to Δ are exactly the horizontal paths which are contained in Σ_Δ. Let us prove this claim. The fact that singular curves are necessary included in Σ_Δ follows by the same argument an in Example 1.14. Let us now prove that any horizontal path which is included in Σ_Δ is singular. Let $\gamma : [0, T] \to M$ such a path be fixed, set $\gamma(0) = x$, and consider a local frame $\{X^1, X^2\}$ for Δ in a neighborhood $\mathcal{V}$ of x. Let $\delta > 0$ be small enough so that $\gamma(t) \in \mathcal{V}$ for any $t \in [0, \delta]$, in such a way that there is $u \in L^2([0, \delta]; \mathbb{R}^2)$ satisfying

$$\dot{\gamma}(t) = u_1(t) X^1(\gamma(t)) + u_2(t) X^2(\gamma(t)) \qquad \text{a.e. } t \in [0, \delta].$$

Taking a change of coordinates if necessary, we can assume that we work in $\mathbb{R}^3$. Let $p_0 \in (\mathbb{R}^3)^* \setminus \{0\}$ be such that $p_0 \cdot X_1(x) = p_0 \cdot X_2(x) = 0$, and let $p : [0, \delta] \to (\mathbb{R}^3)^*$ be the solution to the Cauchy problem

$$\dot{p}(t) = - \sum_{i=1,2} u_i(t) \, p(t) \cdot D_{\gamma(t)} X^i \qquad \text{a.e. } t \in [0, \delta], \quad p(0) = p_0.$$

Define two absolutely continuous function $h_1, h_2 : [0, \delta] \to \mathbb{R}$ by

$$h_i(t) = p(t) \cdot X^i(\gamma(t)) \qquad \forall t \in [0, \delta], \quad \forall i = 1, 2.$$

As above, for every $t \in [0, \delta]$ we have

$$\dot{h}_1(t) = \frac{d}{dt} \Big[p(t) \cdot X^1(\gamma(t)) \Big] = -u_2(t) \, p(t) \cdot [X^1, X^2](\gamma(t))$$

and

$$\dot{h}_2(t) = u_1(t) \, p(t) \cdot \big[X^1, X^2 \big](\gamma(t)).$$

But since $\gamma(t) \in \Sigma_\Delta$ for every t, there are two continuous functions $\lambda_1, \lambda_2 : [0, \delta] \to \mathbb{R}$ such that

$$[X^1, X^2](\gamma(t)) = \lambda_1(t)X^1(\gamma(t)) + \lambda_2(t)X^2(\gamma(t)) \qquad \forall t \in [0, \delta].$$

This implies that the pair (h_1, h_2) is a solution of the linear differential system

$$\begin{cases} \dot{h}_1(t) = -u_2(t)\lambda_1(t)h_1(t) - u_2(t)\lambda_2(t)h_2(t) \\ \dot{h}_2(t) = u_1(t)\lambda_1(t)h_1(t) + u_1(t)\lambda_2(t)h_2(t). \end{cases}$$

Since $h_1(0) = h_2(0) = 0$ by construction, we deduce by the Cauchy-Lipschitz Theorem that $h_1(t) = h_2(t) = 0$ for any $t \in [0, \delta]$. In that way, we have constructed an absolutely continuous arc $p : [0, \delta] \to (\mathbb{R}^3)^* \setminus \{0\}$ satisfying (1.17)–(1.18) (with $\gamma_u = \gamma$). We can repeat this construction on a new interval of the form $[\delta, 2\delta]$ (with initial condition $p(\delta)$) and finally obtain an absolutely continuous arc satisfying (1.17)–(1.18) on $[0, T]$. By Proposition 1.11, we conclude that γ is singular.

Example 1.18 Consider in $\mathbb{R}^4$ the two smooth vector fields X^1, X^2 given by

$$X^1 = \partial_{x_1}, \quad X^2 = \partial_{x_2} + x_1 \partial_{x_3} + x_3 \partial_{x_4}.$$

These two vector fields are always linearly independent in $\mathbb{R}^4$. Moreover we have

$$[X^1, X^2] = \partial_{x_3}, \quad \left[X^2, [X^1, X^2] \right] = -\partial_{x_4}.$$

Therefore the family $\mathscr{F} = \{X^1, X^2\}$ spans a totally nonholonomic distribution Δ of rank two in $\mathbb{R}^4$. Let us look at singular horizontal paths of Δ or equivalently at singular controls with respect to End-Point mapping $E_{\mathscr{F}}^{x,T}$ with $x \in \mathbb{R}^4$ and $T > 0$. Let $u \in U_{\mathscr{F}}^{x,T}$ be a control satisfying $|u(t)| = 1$ for a.e. $t \in [0, T]$. This control is singular if and only if there is an arc $p = (p_1, p_2, p_3, p_4) : [0, T] \to (\mathbb{R}^4)^* \setminus \{0\}$ which satisfies (1.17) and (1.18). Denoting by $x = (x_1, x_2, x_3, x_4) : [0, T] \to \mathbb{R}^4$ the trajectory uniquely associated to x and u, (1.17) yields

$$\begin{cases} \dot{x}_1(t) = u_1(t) \\ \dot{x}_2(t) = u_2(t) \\ \dot{x}_3(t) = u_2(t)x_1(t) \\ \dot{x}_4(t) = u_2(t)x_3(t), \end{cases} \qquad \begin{cases} \dot{p}_1(t) = -u_2(t)p_3(t) \\ \dot{p}_2(t) = 0 \\ \dot{p}_3(t) = -u_2(t)p_4(t) \\ \dot{p}_4(t) = 0, \end{cases} \qquad (1.25)$$

for a.e. $t \in [0, T]$, while (1.18) yields

$$p_1(t) = p_2(t) + x_1(t)p_3(t) + x_3(t)p_4(t) = 0 \qquad \forall t \in [0, T].$$

System (1.25) implies that p_2 and p_4 are constant on $[0, T]$. If $p_4 = 0$, then (1.25) also implies that p_3 is constant on $[0, T]$. Hence we obtain that $p_2 + x_1(t)p_3 = 0$ for every $t \in [0, T]$. Which means that either x_1 is constant or $p_2 = p_3 = 0$. Since p does not vanish on $[0, T]$, we deduce that x_1 is constant, which means that $u_1 \equiv 0$. But $u_2(t)p_3 = 0$ for almost every t, hence $p_3 = 0$ (remember that $|u(t)| = 1$ a.e. $t \in [0, T]$). We obtain a contradiction. Therefore, $p_4 \neq 0$, hence we deduce easily that

$$0 = u_2(t)p_3(t) = \left(-\frac{\dot{p}_3(t)}{p_4}\right)p_3(t) = 0 \qquad \text{a.e. } t \in [0, T].$$

Since p_3 is absolutely continuous, this means that it is constant on $[0, T]$. This implies that $u_2(t) = 0$ for all $t \in [0, T]$. Then, the curve x has the form

$$x(t) = \big(x_1(t), x_2(0), x_3(0), x_4(0)\big) \qquad \forall t \in [0, T].$$

In conclusion, a singular curve passes through each point in $\mathbb{R}^4$.

Example 1.19 The previous phenomena happens for more general rank two distributions in dimension four. Let Δ be a rank two distribution on a four-dimensional manifold M such that for every $x \in M$, there holds

$$\Delta(x) + [\Delta, \Delta](x) \text{ has dimension three}$$

and

$$T_x M = \Delta(x) + [\Delta, \Delta](x) + \big[\Delta, [\Delta, \Delta]\big](x) \qquad \forall x \in M,$$

where

$$\big[\Delta, [\Delta, \Delta]\big](x) := \Big\{[X, [Y, Z]](x) \,|\, X, Y, Z \text{ sections of } \Delta\Big\}.$$

As above, we can work locally, so let us consider a frame $\{X^1, X^2\}$ and a trajectory $x : [0, T] \to \mathbb{R}^4$ associated to some control $u \in L^2([0, T]; \mathbb{R}^2)$. If x is singular (with respect to Δ), there is $p : [0, T] \to (\mathbb{R}^4)^* \setminus \{0\}$ satisfying (1.17) and (1.18). Derivativing (1.18) two times yields for almost every $t \in [0, T]$ with $u(t) \neq 0$,

$$p(t) \cdot [X^1, X^2](x(t)) = 0 \quad \text{and} \tag{1.26}$$

$$u_1(t)\, p(t) \cdot \big[X^1, [X^1, X^2]\big](x(t)) + u_2(t)\, p(t) \cdot \big[X^2, [X^1, X^2]\big](x(t)) = 0. \tag{1.27}$$

Since M has dimension four and $\Delta + [\Delta, \Delta]]$ has dimension three, there is (locally) a smooth non-vanishing 1-form α whose kernel is equal to $\Delta + [\Delta, \Delta]$. Then, by (1.18) and (1.26) α and $p(t)$ are colinear along x, and in turn by (1.27) we have for almost every $t \in [0, T]$ with $u(t) \neq 0$,

$$u_1(t)\,\alpha_{x(t)} \cdot \left[X^1, \left[X^1, X^2\right]\right](x(t)) + u_2(t)\,\alpha_{x(t)} \cdot \left[X^2, \left[X^1, X^2\right]\right](x(t)) = 0.$$

By the above assumptions, for every x, the linear form

$$(\lambda_1, \lambda_2) \in \mathbb{R}^2 \longmapsto \left(\alpha_x \cdot \left[X^1, \left[X^1, X^2\right]\right](x)\right)\lambda_1 + \left(\alpha_x \cdot \left[X^2, \left[X^1, X^2\right]\right](x)\right)\lambda_2$$

has a kernel of dimension one. This shows that there is a smooth line field (a distribution of rank one) $L \subset \Delta$ on M such that the singular curves are exactly the integral curves of L.

1.4 The Chow-Rashevsky Theorem

Openness of End-Point mappings. The following result will imply easily the Chow-Rashevsky Theorem. We recall that a map is said to be open if the image of any open set is open.

Proposition 1.12 *Let $\mathscr{F} = \left\{X^1, \ldots, X^k\right\}$ be a family of smooth vector fields on M satisfying the Hörmander condition on M. Then for every $x \in M$ and every $T > 0$, the End-Point mapping $E_{\mathscr{F}}^{x,T} : U_{\mathscr{F}}^{x,T} \to M$ is open.*

Proof Let $x \in M$ and $T > 0$ be fixed. Set for every $\varepsilon > 0$,

$$d(\varepsilon) = \max\left\{\mathrm{rank}_{\mathscr{F}}^{x,\varepsilon}(u) \mid u \in U_{\mathscr{F}}^{x,\varepsilon} \text{ s.t. } \|u\|_{L^2} < \varepsilon\right\}.$$

By Proposition 1.9 (iii), the function $\varepsilon \in (0, +\infty) \mapsto d(\varepsilon)$ is nondecreasing with values in $\mathbb{N}$. So, there is ε_0 and $d_0 \in \mathbb{N}$ such that $d(\varepsilon) = d_0$ for any $\varepsilon \in (0, \varepsilon_0)$. Since $\mathscr{F}$ satisfies the Hörmander condition at x, the vector space spanned by $\{X^1(x), \ldots, X^k(x)\}$ has dimension ≥ 1. Then, thanks to Proposition 1.10, there holds

$$d(\varepsilon) = d_0 \geq 1 \qquad \forall \varepsilon \in [0, \varepsilon_0].$$

Let $\varepsilon \in (0, \varepsilon_0)$ and $u^\varepsilon \in U_{\mathscr{F}}^{x,\varepsilon}$ such that $\|u^\varepsilon\|_{L^2} < \varepsilon$ and $\mathrm{rank}_{\mathscr{F}}^{x,\varepsilon}(u^\varepsilon) = d_0$ be fixed. There are d_0 controls $v^1, \ldots, v^{d_0}$ in $L^2([0, \varepsilon]; \mathbb{R}^k)$ such that the linear map

$$\mathscr{L} : \mathbb{R}^{d_0} \longrightarrow T_x M$$
$$\lambda = (\lambda^1, \ldots, \lambda^{d_0}) \longmapsto D_{u^\varepsilon} E_{\mathscr{F}}^{x,\varepsilon}\left(\sum_{j=1}^{d_0} \lambda^j v^j\right)$$

is injective. By construction and the fact that the mapping $u \mapsto \mathrm{rank}_{\mathscr{F}}^{x,\varepsilon}(u)$ is lower semicontinuous, the rank of any control u is equal to $\mathrm{rank}_{\mathscr{F}}^{x,\varepsilon}(u^\varepsilon) = d_0$ as soon as u is close enough to u^ε in $L^2([0, \varepsilon]; \mathbb{R}^k)$. Hence, there is an open neighborhood $\mathscr{V}$ of $0 \in \mathbb{R}^{d_0}$ where the mapping

$$\mathcal{E} : \mathcal{V} \longrightarrow M$$
$$\lambda \longmapsto E^{x,\varepsilon}_{\mathcal{F}}\left(u^{\varepsilon} + \sum_{j=1}^{d_0} \lambda^j v^j\right)$$

is an embedding whose image is a submanifold N of class C^1 in M of dimension d_0. Moreover by construction again, there holds for every small $\lambda \in \mathbb{R}^{d_0}$,

$$\mathrm{Im}^{x,\varepsilon}_{\mathcal{F}}\left(u^{\varepsilon} + \sum_{j=1}^{d_0} \lambda^j v^j\right) = D_\lambda \mathcal{E}\left(\mathbb{R}^{d_0}\right) = T_{\mathcal{E}(\lambda)}N.$$

By Proposition 1.10, we infer that $X^i(y)$ belongs to $T_y N$ for any $i = 1, \ldots, k$ and $y \in N$.

Lemma 1.13 *Let Ω be an open subset of $\mathbb{R}^l$ ($l \geq 2$) and $\mathcal{S}$ be a submanifold of Ω of class C^1. Let X, Y be two smooth vector fields on Ω such that*

$$X(x), Y(x) \in T_x \mathcal{S} \quad \forall x \in \mathcal{S}.$$

Then $[X, Y](x) \in T_x \mathcal{S}$ for any $x \in \mathcal{S}$.

Proof (Proof of Lemma 1.13) As in Proposition 1.5, we denote respectively by e^{tX} and e^{tY} the flows of X and Y. Since by assumption X and Y is always tangent to $\mathcal{S}$, $e^{tX}(x)$ and $e^{tY}(x)$ belong to $\mathcal{S}$ for any $x \in \mathcal{S}$ and any t small. Therefore

$$\left(e^{-tY} \circ e^{-tX} \circ e^{tY} \circ e^{tX}\right)(x) \in \mathcal{S} \quad \forall x \in \mathcal{S} \text{ and } t \text{ small.}$$

By Proposition 1.5, we infer that $[X, Y](x) \in T_x \mathcal{S}$ for any $x \in \mathcal{S}$. $\qquad \square$

From the above lemma it follows that all the brackets involving $X^1, \ldots, X^k$ at $y \in N$ belong to $T_y N$. Since $\mathcal{F}$ satisfies the Hörmander condition, this shows that $d_0 = n$ and indeed that $d(\varepsilon) = n$ for any $\varepsilon > 0$.

Let $\mathcal{O}$ be an open subset of $U^{x,T}_{\mathcal{F}}$, $v \in \mathcal{O}$ and $\varepsilon > 0$ to be chosen later. Since $d(\varepsilon) = n$, there is $u \in U^{x,\varepsilon}_{\mathcal{F}}$ such that $\|u\|_{L^2} < \varepsilon$ and $\mathrm{rank}^{x,\varepsilon}_{\mathcal{F}}(u) = n$. Define the control $\tilde{v} \in L^2([0, T]; \mathbb{R}^k)$ by

$$\tilde{v} = u * \check{u} * v_{\frac{T}{T-2\varepsilon}}.$$

The trajectory associated with $\tilde{v}$ is the concatenation of the curve x_u starting at x and associated with u, $x_{\check{u}}$ starting at $x_u(\varepsilon)$ and associated with $\check{u}$, and a reparametrization of x_v starting at x and associated with v (see Fig. 1.4).

By construction (see Proposition 1.9), $\tilde{v}$ belongs to $U^{x,T}_{\mathcal{F}}$, is regular and satisfies $E^{x,T}_{\mathcal{F}}(\tilde{v}) = E^{x,T}_{\mathcal{F}}(v)$. Then, as above, the image of a small ball centered at the origin in $L^2([0, T]; \mathbb{R}^k)$ by the mapping $w \mapsto E^{x,T}_{\mathcal{F}}(\tilde{v} + w)$ is an open neighboorhood of $E^{x,T}_{\mathcal{F}}(\tilde{v}) = E^{x,T}_{\mathcal{F}}(v)$. Furthermore, the quantity $\left\|\tilde{v} - v\right\|^2_{L^2}$ is equal to

Fig. 1.4 The trajectory associated with $\tilde{v}$

$$\int_0^\varepsilon |u(t) - v(t)|^2 \, dt + \int_\varepsilon^{2\varepsilon} |-u(2\varepsilon - t) - v(t)|^2 \, dt$$

$$+ \int_{2\varepsilon}^T \left| \left(\frac{T}{T - 2\varepsilon} \right) v \left(\frac{T(t - 2\varepsilon)}{T - 2\varepsilon} \right) - v(t) \right|^2 \, dt$$

and consequently bounded by

$$2 \int_0^\varepsilon |u(t)|^2 \, dt + \int_0^{2\varepsilon} |v(t)|^2 \, dt - 2 \int_0^\varepsilon \langle u(t), v(t) \rangle \, dt + 2 \int_\varepsilon^{2\varepsilon} \langle u(2\varepsilon - t), v(t) \rangle \, dt$$

$$+ \int_{2\varepsilon}^T \left| \left(\frac{T}{T - 2\varepsilon} \right) u(t) - u(t) \right|^2 \, dt + \left(\frac{T}{T - 2\varepsilon} \right)^2 \int_{2\varepsilon}^T \left| u \left(\frac{T(t - 2\varepsilon)}{T - 2\varepsilon} \right) - u(t) \right|^2 \, dt$$

$$+ 2 \left(\frac{T}{T - 2\varepsilon} \right) \int_{2\varepsilon}^T \left\langle u \left(\frac{T(t - 2\varepsilon)}{T - 2\varepsilon} \right) - u(t), \left(\frac{T}{T - 2\varepsilon} \right) u(t) - u(t) \right\rangle \, dt.$$

Then since $\|u\|_{L^2} < \varepsilon$ and both functions

$$t \in [2\varepsilon, T] \quad \longmapsto \quad \left(\frac{T}{T - 2\varepsilon} \right) u(t) - u(t)$$

$$\text{and} \quad t \in [2\varepsilon, T] \quad \longmapsto \quad u \left(\frac{T(t - 2\varepsilon)}{T - 2\varepsilon} \right) - u(t)$$

tend to zero in L^2, we infer that $\tilde{v}$ belong to $\mathcal{O}$ if ε is small enough. This shows that $E_{\mathscr{F}}^{x,T}(\mathcal{O})$ contains a neighborhood of $E_{\mathscr{F}}^{x,T}(\tilde{v}) = E_{\mathscr{F}}^{x,T}(v)$. $\qquad \square$

Statement and proof. The aim of the present section is to prove the following result.

Theorem 1.14 (Chow-Rashevsky's Theorem) *Let Δ be a totally nonholonomic distribution on M (assumed to be connected). Then, for every $x, y \in M$ and every $T > 0$, there is an horizontal path $\gamma \in \Omega_\Delta^{x,T}$ such that $\gamma(T) = y$.*

Thanks to the above discussion, the Chow-Rashevsky Theorem will be a straightforward consequence of the following result.

Theorem 1.15 *Let $\mathscr{F} = \{X^1, \ldots, X^k\}$ be a family of smooth vector fields on M. Assume that M is connected and that $\mathscr{F}$ satisfies the Hörmander condition on M. Then for every $x, y \in M$ and every $T > 0$, there is a control $u \in U_{\mathscr{F}}^{x,T}$ such that the solution of (1.6) satisfies $\gamma_u(T) = y$.*

Fig. 1.5 Proof of the Chow-Rashevsky Theorem

Proof Let x and $T > 0$ be fixed. Denote by $\mathscr{A}_{\mathscr{F}}(x, T)$ the set of points in M which can be joined from x by a control in $U_{\mathscr{F}}^{x,T}$, that is

$$\mathscr{A}_{\mathscr{F}}(x, T) = E_{\mathscr{F}}^{x,T}\left(U_{\mathscr{F}}^{x,T}\right).$$

By Proposition 1.12, $\mathscr{A}_{\mathscr{F}}(x, T)$ is an open set in M. Let us show that this set is closed as well. Let $\{z_k\}_k$ be a sequence of points in M converging to some $z \in M$. By openness of the mapping $E_{\mathscr{F}}^{z,1}$ and the fact that $E_{\mathscr{F}}^{z,1}(0) = z$, the set $E_{\mathscr{F}}^{z,1}\left(U_{\mathscr{F}}^{z,1}\right)$ is an neighborhood of z. Then, there is k large enough such that z_k belongs to that set.

The concatenation of u_k together with $\breve{u}$ steers x to z (see Fig. 1.5). This shows that $\mathscr{A}_{\mathscr{F}}(x, T)$ is closed in M. In conclusion $\mathscr{A}_{\mathscr{F}}(x, T)$ is open, closed and nonempty (it contains x). By connectedness of M, we infer that $\mathscr{A}_{\mathscr{F}}(x, T) = M$. □

Remark 1.10 The Chow-Rashevsky may be of course obtained in different ways. For instance, consider in $\mathbb{R}^3$ a totally nonholonomic rank two distribution Δ generated by two smooth vector fields X^1, X^2 such that

$$\text{Span}\left\{X^1(x), X^2(x), [X^1, X^2](x)\right\} = \mathbb{R}^3 \quad \forall x \in \mathbb{R}^3.$$

Let $x \in \mathbb{R}^3$ and $\lambda > 0$ be fixed, define the function $\Phi_\lambda : \mathbb{R}^3 \to \mathbb{R}^3$ by

$$\Phi_\lambda(t_1, t_2, t_3) := \left(e^{\lambda X^1} \circ e^{t_3 X^2} \circ e^{-\lambda X^1} \circ e^{t_2 X^2} \circ e^{t_1 X^1}\right)(x),$$

for every $(t_1, t_2, t_3) \in \mathbb{R}^3$. It can be shown that Φ_λ is a local diffeomorphism in a neighborhood of the origin provided λ is small enough. This implies easily the Chow-Rashevsky Theorem for contact distributions in dimension three.

1.5 Sub-Riemannian Structures

Definition. A *sub-Riemannian structure* on M is given by a pair (Δ, g) where Δ is a totally nonholonomic distribution on M and g is a smooth Riemannian metric on Δ, that is for every $x \in M$, $g(\cdot, \cdot)$ is a scalar product on Δ_x. A simple way to construct a sub-Riemannian structure is to take a smooth connected Riemannian manifold (M, g), to consider a totally nonholonomic distribution Δ on M, and to take as sub-Riemannian metric the restriction of g to the distribution. In fact, any sub-Riemannian structure can be obtained in this way.

Example 1.20 The space $\mathbb{R}^3$ (with coordinates (x, y, z)) equipped with the rank two distribution Δ given in Example 1.2 and with the metric $g = dx^2 + dy^2$ is the most simple sub-Riemannian structure we can imagine. The length of a vector $v = (v_1, v_2, v_3) \in \Delta(x, y, z)$ is given by

$$|v|^g = \sqrt{v_1^2 + v_2^2}.$$

Since v is an horizontal vector, the latter quantity does not vanishes unless $v = 0$.

If the distribution Δ admits a frame $X^1, \ldots, X^m$ on an open set $\mathcal{O} \subset M$, then the family $\mathcal{F} = \{X^1, \ldots, X^m\}$ is called an *orthonormal dgenerating family of vector fields* or an *orthonormal frame* for (Δ, g) in $\mathcal{O}$ if there holds

$$g_x\left(X^i(x), X^j(x)\right) = \delta_{ij} \qquad \forall i, j = 1, \ldots, m, \quad \forall x \in \mathcal{O},$$

where δ_{ij} denotes the Kronecker symbol (that is $\delta_{ij} = 1$ if $i = j$ and $\delta_{ij} = 0$ if $i \neq j$). Sub-Riemannian structures admit local orthonormal frames in a neighborhood of each point of M.

The sub-Riemannian distance. From now on, for every $x \in M$ we denote by $|\cdot|_x^g$ the *sub-Riemannian norm* on $\Delta(x)$, that is

$$|v|_x^g = \sqrt{g_x(v, v)} \qquad \forall v \in \Delta(x).$$

The *length* of an horizontal path $\gamma \in \Omega_\Delta^{x,T}$ is defined by

$$\text{length}^g(\gamma) := \int_0^T |\dot\gamma(t)|_{\gamma(t)}^g dt.$$

Note that since any horizontal path is absolutely continuous with square integrable derivative, the length of any horizontal path is finite.

Let (Δ, g) be a sub-Riemannian structure on M, by the Chow-Rashevsky Theorem, for every $x, y \in M$, there is at least one horizontal path joining x to y in time 1. For every $x, y \in M$, the *sub-Riemannian distance* between x and y, denoted by $d_{SR}(x, y)$, is defined as the infimum of lengths of horizontal paths joining x to y, that is,

$$d_{SR}(x, y) := \inf\left\{\text{length}^g(\gamma) \mid \gamma \in \Omega_\Delta^{x,1} \text{ s.t. } \gamma(1) = y\right\}.$$

The function d_{SR} defines a distance on $M \times M$ (the triangular inequality is easy, the fact that $d_{SR}(x, y) \Rightarrow x = y$ follows from the proof of Proposition 1.11) and makes M a metric space. Given $x \in M$ and $r \geq 0$, we call *sub-Riemannian ball* centered at x with radius r the set defined as

$$B_{SR}(x, r) = \left\{y \in M \mid d_{SR}(x, y) < r\right\}.$$

The openness of End-Point mappings associated with totally nonholonomic distribution yields the following result.

Proposition 1.16 *Let (Δ, g) be a sub-Riemannian structure on M, then the topology defined by d_{SR} coincides with the original topology of M. In particular, the sub-Riemannian distance d_{SR} is continuous on $M \times M$.*

Proof We need to show that for every $x \in M$, the family of sub-Riemannian balls $\{B_{SR}(x, r)\}_{r>0}$ is a basis of neighborhoods for x with respect to the original topology. Let $\mathscr{F} = \{X^1, \ldots, X^m\}$ be an orthonormal frame for Δ on an open neighborhood $\mathscr{V}_x$ of some $x \in M$ Let $\mathscr{V} \subset \mathscr{V}_x$ be an open and relatively compact neighborhood of x with respect to the initial topology. Let us show that there is $r > 0$ small enough such that $B_{SR}(x, r) \subset \mathscr{V}$. Let $\mathscr{W}$ be an open neighborhood of x such that $\overline{\mathscr{W}} \subset \mathscr{V}$. Define the compact annulus $\mathscr{A}$ by

$$\mathscr{A} = \overline{\mathscr{V}} \setminus \mathscr{W}.$$

Any continuous path joining x to a point outside $\mathscr{V}$ has to cross $\mathscr{A}$. Hence since $X^1, \ldots, X^m$ are bounded on $\mathscr{A}$, there is $\delta > 0$ such that any solution $\gamma_u : [0, 1] \to M$ to the Cauchy problem

$$\dot{\gamma}_u(t) = \sum_{i=1}^{m} u_i(t) X^i (\gamma_u(t)) \qquad \text{a.e. } t \in [0, 1], \quad \gamma_u(0) = x$$

with $u \in U_{\mathscr{F}}^{x,1}$ and $\gamma_u(1) \notin \mathscr{V}$ satisfies

$$\int_0^1 \left| \sum_{i=1}^{m} u_i(t) X^i (\gamma_u(t)) \right|_{\gamma_u(t)}^{g} dt > \delta.$$

By Proposition 1.7, this means that the sub-Riemannian ball $B_{SR}(x, \delta/2)$ is included in $\mathscr{V}$. Let us now show that any sub-Riemannian ball $B_{SR}(x, r)$ contains an open neighborhood of x with respect to the initial topology. The set $U_{\mathscr{F}}^{x,1}$ is open in $L^2([0, 1]; \mathbb{R}^m)$ and contains the control $u \equiv 0$. Thus there is $\nu > 0$ such that the L^2-ball $B_{L^2}(0, \nu)$ is contained in $U_{\mathscr{F}}^{x,1}$. Moreover, since $\mathscr{F}$ is orthonormal with respect to g, there holds for every $u \in B_{L^2}(0, \nu)$

$$\text{length}^g(\gamma_u) = \int_0^1 |u(t)| dt.$$

Thanks to the Cauchy-Schwarz inequality, we infer that

$$E_{\mathscr{F}}^{x,1} (B_{L^2}(0, \nu)) \subset B_{SR}(x, \nu).$$

Proposition 1.12 together with $E_{\mathscr{F}}^{x,1} = x$ concludes the proof. $\qquad\square$

1.6 Notes and Comments

Proposition 1.3 is taken from a paper by Sussmann [11]. The Hörmander condition introduced in Sect. 1.1 is also refered as bracket generating condition. The term comes from the analysis literature; it is named after Hörmander who obtained hypoellipticity results for linear operators associated with families of vector fields [6]. Several other terms may be used to refer to totally nonholonomic distributions. They are called bracket generating by Montgomery [8], nonholonomic by Bellaiche [2], and they refer to completely nonholonomic families of vector fields by Agrachev and Sachkov [1].

The notion of singular curves play a major role in this monograph. Most of the examples of singular horizontal paths given in Sect. 1.3 are classical. The most valuable (Example 1.19) is taken from [10].

Theorem 1.14 has been proved independently by Chow [4] and Rashevsky [9] in the 1930s, see [4, 9]. The proof that we present here is an adaptation of the one given by Bellaiche [2] to prove the so-called Orbit Theorem (see also [1, 7]). Other proofs of the Chow-Rashevsky Theorem can be found in the texts of Bismut [3], Gromov [5], or Montgomery [8].

References

1. Agrachev, A.A., Sachkov, Yu.L.: Control Theory from the Geometric Viewpoint. Encyclopaedia of Mathematical Sciences, vol. 87. Springer-Verlag, Heidelberg (2004)
2. Bellaiche, A.: The tangent space in sub-Riemannian geometry. In: Sub-Riemannian Geometry, pp. 1–78. Birkhäuser, Basel (1996)
3. Bismut, J.-M.: Large Deviations and the Malliavin Calculus. Progress in Mathematics, vol. 45. Birkhäuser, Boston (1984)
4. Chow, C.-L.: Über systeme von linearen partiellen differentialgleichungen ester ordnung. Math. Ann. **117**, 98–105 (1939)
5. Gromov, M.: Carnot-Carathéodory spaces seen from within. In: Sub-Riemannian Geometry, pp. 79–323. Birkhäuser, Basel (1996)
6. Hörmander, L.: Hypoelliptic second order differential operators. Acta. Math. **119**, 147–171 (1967)
7. Jurdjevic, V.: Geometric Control Theory. Cambridge Studies in Advanced Mathematics, vol. 52. Cambridge University Press, Cambridge (1997)
8. Montgomery, R.: A tour of subriemannian geometries, their geodesics and applications. In: Mathematical Surveys and Monographs, vol. 91. American Mathematical Society, Providence, RI (2002)
9. Rashevsky, P.K.: About connecting two points of a completely nonholonomic space by admissible curve. Uch. Zapiski Ped. Inst. Libknechta **2**, 83–94 (1938)
10. Sussmann, H.J.: A cornucopia of four-dimensional abnormal sub-Riemannian minimizers. In: Sub-Riemannian Geometry, pp. 341–364. Birkhäuser, Basel (1996)
11. Sussmann, H.J.: Smooth distributions are globally finitely spanned. In: Analysis and Design of Nonlinear Control Systems, pp. 3–8. Springer-Verlag, Heidelberg (2008)

Chapter 2
Sub-Riemannian Geodesics

Throughout all the chapter, M denotes a smooth connected manifold without boundary of dimension $n \geq 2$ equipped with a sub-Riemannian structure (Δ, g) of rank $m \leq n$.

2.1 Minimizing Horizontal Paths and Geodesics

Definition. Given $x, y \in M$, we call *minimizing horizontal path* between x and y any path $\gamma \in \Omega_\Delta^{x,T}$ with $T \geq 0$ such that

$$d_{SR}(x, y) = \text{length}^g(\gamma).$$

Like in the Riemannian case, minimizing paths with constant speed minimize the so-called sub-Riemannian energy. Given $x, y \in M$, we define the *sub-Riemannian energy* between x and y by

$$e_{SR}(x, y) := \inf \left\{ \text{energy}^g(\gamma) \mid \gamma \in \Omega_\Delta^{x,1} \text{ s.t. } \gamma(1) = y \right\},$$

where the *energy* of a path $\gamma \in \Omega_\Delta^{x,1}$ is defined as

$$\text{energy}^g(\gamma) := \int_0^1 \left(|\dot{\gamma}(t)|_{\gamma(t)}^g \right)^2 dt.$$

The following result whose proof is based on Cauchy-Schwarz's inequality, is fundamental.

Proposition 2.1 *For any* $x, y \in M$, $e_{SR}(x, y) = d_{SR}(x, y)^2$.

L. Rifford, *Sub-Riemannian Geometry and Optimal Transport*,
SpringerBriefs in Mathematics, DOI: 10.1007/978-3-319-04804-8_2,
© The Author(s) 2014

Proof Let $x, y \in M$ be fixed. First, we observe that, for every horizontal path γ : $[0, 1] \to M$ satisfying $\gamma(0) = x$ and $\gamma(1) = y$, the Cauchy-Schwarz inequality yields

$$\left(\int_0^1 |\dot{\gamma}(t)|^g_{\gamma(t)} \, dt \right)^2 \leq \int_0^1 \left(|\dot{\gamma}(t)|^g_{\gamma(t)} \right)^2 dt. \tag{2.1}$$

Taking the infimum over the set of $\gamma \in \Omega_\Delta^{x,1}$ such that $\gamma(1) = y$ yields $d_{SR}(x, y)^2 \leq e_{SR}(x, y)$. On the other hand, for every $\varepsilon > 0$, there exists an horizontal path $\gamma \in \Omega_\Delta^{x,1}$, with $\gamma(1) = y$, such that

$$\text{length}^g(\gamma) = \int_0^1 |\dot{\gamma}(t)|^g_{\gamma(t)} \, dt \leq d_{SR}(x, y) + \varepsilon.$$

Reparametrizing γ by arc-length, we get a new path $\xi \in \Omega_\Delta^{x,1}$ with $\gamma(1) = y$ satisfying

$$\left| \dot{\xi}(t) \right|^g_{\xi(t)} = \text{length}^g(\gamma) \qquad \text{a.e. } t \in [0, 1].$$

Consequently,

$$e_{SR}(x, y) \leq \int_0^1 \left(|\dot{\xi}(t)|^g_{\xi(t)} \right)^2 dt = \text{length}^g(\gamma)^2 \leq (d_{SR}(x, y) + \varepsilon)^2.$$

Letting ε tend to 0 completes the proof of the result.   □

Given $x, y \in M$, we call *minimizing geodesic* between x and y any path $\gamma \in \Omega_\Delta^{x,1}$ joining x to y such that

$$e_{SR}(x, y) = \text{energy}^g(\gamma).$$

Thanks to the above proof and the fact that equality holds in the Cauchy-Schwarz inequality (2.1) if and only γ has constant speed (that is $|\dot{\gamma}(t)|^g_{\gamma(t)}$ is constant), we obtain the following result.

Proposition 2.2 *Given $x, y \in M$, a path $\gamma \in \Omega_\Delta^{x,1}$ is a minimizing geodesic between x and y if and only if it is a minimizing horizontal path between x and y with constant speed.*

Sufficiently near points can be joined by minimizing geodesics and a fortiori by minimizing horizontal paths.

Proposition 2.3 *Let $x \in M$, then there is $\rho > 0$ such that the following property is satisfied:*
For every $y, z \in B_{SR}(x, \rho)$ and any minimizing sequence $\{\gamma^k\}_k : [0, 1] \to M$ of horizontal paths with constant speed such that

$$\lim_{k \to +\infty} \gamma^k(0) = y, \quad \lim_{k \to +\infty} \gamma^k(1) = z, \quad \lim_{k \to +\infty} \text{length}^g(\gamma^k) = d_{SR}(y, z), \tag{2.2}$$

up to taking a subsequence, $\{\gamma^k\}_k$ converges uniformly to some minimizing geodesic $\bar{\gamma} \in \Omega_{\Delta}^{y,1}$ joining y to z.

In particular, for every $y, z \in B_{SR}(x, \rho)$, there is a minimizing geodesic between y and z.

Proof Fix $x \in M$ and $\mathscr{F} = \{X^1, \dots, X^m\}$ an orthonormal frame for Δ on an open and relatively compact neighborhood $\mathscr{V}_x$ of x. From Proposition 1.13, there is $r > 0$ small enough such that $B_{SR}(x, r) \subset \mathscr{V}_x$. For any $y \in B_{SR}(x, r/4)$ and any horizontal path $\gamma : [0, 1] \to M$ with constant speed satisfying

$$d_{SR}(\gamma(0), y) < \frac{r}{24} \quad \text{and} \quad \text{length}^g(\gamma) \leq \frac{2r}{3},$$

we have for every $t \in [0, 1]$,

$$\begin{aligned} d_{SR}(x, \gamma(t)) &\leq d_{SR}(x, y) + d_{SR}(y, \gamma(t)) \\ &\leq r/4 + d_{SR}(y, \gamma(0)) + \text{length}^g(\gamma) \\ &\leq r/4 + r/24 + 2r/3 = 23r/24 < r. \end{aligned}$$

Which means that γ is contained in $B_{SR}(x, r)$. Furthermore, for every such horizontal path, there is $u \in L^2([0, 1]; \mathbb{R}^m)$ such that

$$\dot{\gamma}(t) = \sum_{i=1}^m u_i(t) X^i(\gamma(t)) \quad \text{and} \quad |\dot{\gamma}(t)|_{\gamma(t)}^g = \|u\|_{L^2} = \text{length}^g(\gamma),$$

for a.e. $t \in [0, 1]$. Let $y, z \in B_{SR}(x, r/4)$ be fixed and $\{\gamma^k\}_k : [0, 1] \to M$ be a sequence of horizontal paths with constant speed verifying (2.2). By the above discussion, we may assume without loss of generality that all the paths $\gamma^k : [0, 1] \to M$ are valued in the compact set $\overline{\mathscr{V}_x}$ with derivatives bounded by r and associated with a sequence of controls $\{u^k\}_k$ in $L^2([0, 1]; \mathbb{R}^m)$ such that $\|u^k\|_{L^2} = \text{length}^g(\gamma^k)$. Then by Arzela-Ascoli's theorem taking a subsequence if necessary the sequence $\{\gamma^k\}_k$ converges to some $\bar{\gamma} : [0, 1] \to M$. Moreover, the sequence $\{u^k\}_k$ is bounded in L^2 so it weakly converges up to a subsequence to some $\bar{v} \in L^2([0, 1]; \mathbb{R}^m)$. We obtain easily that $\bar{\gamma}(0) = y, \bar{\gamma}(1) = z$,

$$\dot{\bar{\gamma}}(t) = \sum_{i=1}^m \bar{v}_i(t) X^i(\bar{\gamma}(t)) \quad \text{a.e. } t \in [0, 1],$$

and by lower semicontinuity of the L^2-norm under weak convergence we immediately deduce that

$$\|\bar{v}\|_{L^2} \leq \lim_{k \to +\infty} \left\|u^k\right\|_{L^2} = d_{SR}(y, z).$$

Furthermore, since $\bar{\gamma}$ is an horizontal path joining y to z, there holds

$$d_{SR}(y, z) \leq \text{length}^g(\bar{\gamma}).$$

By Cauchy-Schwarz's inequality, we have $\text{length}^g(\bar{\gamma}) \leq \|\bar{v}\|_{L^2}$. Then we infer that

$$\text{energy}^g(\bar{\gamma}) = \|\bar{v}\|_{L^2}^2 = d_{SR}(y, z)^2 = e_{SR}(y, z).$$

Which shows that $\bar{\gamma}$ is a minimizing geodesic joining y to z.   □

Remark 2.1 The above proof shows indeed that up to taking a subsequence, the sequence $\{u^k\}_k$ converges strongly to $\bar{v}$ in $L^2([0, 1]; \mathbb{R}^m)$. As a matter of fact, it converges weakly to $\bar{v}$ and satisfies

$$\lim_{k \to +\infty} \left\|u^k\right\|_{L^2} = \|\bar{v}\|_{L^2}.$$

The SR Hopf-Rinow Theorem. The following sub-Riemannian version of the classical Riemannian Hopf-Rinow Theorem holds.

Theorem 2.4 (Hopf-Rinow Theorem) *Let (Δ, g) be a sub-Riemannian structure on M. Assume that (M, d_{SR}) is a complete metric space. Then the following properties hold:*

(i)   The balls $\bar{B}_{SR}(x, r)$ are compact (fort any $r \geq 0$).
(ii) For every $x, y \in M$ there exists at least one minimizing geodesic joining x to y.

Proof Let us first recall that thanks to Proposition 1.13, the metric space (M, d_{SR}) is locally compact. That is for every $x \in M$, there is $r > 0$ such that the ball $\bar{B}_{SR}(x, r)$ is compact. Let $x \in M$ be fixed. We first show that all the balls $\bar{B}_{SR}(x, r)$ with $r \geq 0$ are compact. Denote by I_x the set of $r \geq 0$ such that $\bar{B}_{SR}(x, r)$ is compact. By inclusion of the balls $\bar{B}_{SR}(x, r') \subset \bar{B}_{SR}(x, r)$ if $r' \leq r$ and local compactness of (M, d_{SR}), I_x is an interval whose supremum R_x is strictly positive. We claim that I is both closed and open in $[0, +\infty)$.

Lemma 2.5 *The interval I_x is closed in $[0, +\infty)$.*

Proof (Proof of Lemma 2.5) We need to show that R_x belongs to I_x, that is that $\bar{B}_{SR}(x, R_x)$ is compact. Let $\{y_k\}_k$ be a sequence of points in $\bar{B}_{SR}(x, R_x)$, we need to show that it has a convergent subsequence. We construct a Cauchy subsequence of $\{y_k\}_k$ as follows. For every integer $l \geq 1$, we set

$$K^l = \bar{B}_{SR}\left(x, R_x\left(1 - 2^{-l}\right)\right).$$

By assumption, $\{K^l\}_l$ is an increasing sequence of compact sets in $\bar{B}_{SR}(x, R_x)$. For every $k \in \mathbb{N}$, there is $y_k^l \in K^1$ such that

$$d_{SR}\left(y_k, y_k^1\right) = \inf\left\{d_{SR}(y_k, z) \mid z \in K^1\right\} \leq \frac{R_x}{2}.$$

By compactness of K^1, there is a strictly increasing mapping $\varphi^1 : \mathbb{N} \to \mathbb{N}$ such that the sequence $\{y_{\varphi^1(k)}^1\}_k$ converges to some $\bar{y}^1 \in K^1$. Thus there exists $k_1 \geq 0$ such that

$$d_{SR}\left(y_{\varphi^1(k)}^1, \bar{y}^1\right) \leq \frac{R_x}{2} \qquad \forall k \geq k_1.$$

Set $z_1 := y_{\varphi^1(k_1)}$. Now for every $k \in \mathbb{N}$, there is $y_k^2 \in K^2$ such that

$$d_{SR}\left(y_{\varphi^1(k)}, y_k^2\right) = \inf\left\{d_{SR}(y_{\varphi^1(k)}, z) \mid z \in K^2\right\} \leq \frac{R_x}{4}.$$

Again, by compactness of K^2 there exists a strictly increasing mapping $\varphi^2 : \mathbb{N} \to \mathbb{N}$ such that the sequence $\{y_{\varphi^2(k)}^2\}_k$ converges to some $\bar{y}^2 \in K^2$ and then there is $k_2 \geq k_1$ such that

$$d_{SR}\left(y_{\varphi^2(k)}^2, \bar{y}^2\right) \leq \frac{R_x}{4} \qquad \forall k \geq k_2.$$

Set $z_2 := y_{(\varphi^1 \circ \varphi^2)(k_2)}$. By construction, there holds

$$
\begin{aligned}
d_{SR}(z_1, z_2) &\leq d_{SR}\left(z_1, y_{\varphi^1(k_1)}^1\right) + d_{SR}\left(y_{\varphi^1(k_1)}^1, z_2\right) \\
&= d_{SR}\left(y_{\varphi^1(k_1)}, y_{\varphi^1(k_1)}^1\right) + d_{SR}\left(y_{\varphi^1(k_1)}^1, y_{(\varphi^1 \circ \varphi^2)(k_2)}\right) \\
&\leq \frac{R_x}{2} + d_{SR}\left(y_{\varphi^1(k_1)}^1, \bar{y}^1\right) + d_{SR}\left(\bar{y}^1, y_{(\varphi^1 \circ \varphi^2)(k_2)}^1\right) + d_{SR}\left(y_{(\varphi^1 \circ \varphi^2)(k_2)}^1, z_2\right) \\
&\leq \frac{R_x}{2} + \frac{R_x}{2} + \frac{R_x}{2} + \frac{R_x}{2} \leq 2R_x.
\end{aligned}
$$

Repeating this construction yields a sequence of strictly increasing mappings $\{\varphi^l\}_l$, a sequence (with two indices) $\{y_k^l\}_{k,l}$, a sequence of limits $\{\bar{y}^l\}_l$, and a nondecreasing sequence of integers $\{k_l\}_l$ such that

$$d_{SR}\left(y_k, y_k^l\right) = \inf\left\{d_{SR}(y_k, z) \mid z \in K^l\right\} \leq \frac{R_x}{2^l}$$

and

$$d_{SR}\left(y_{\varphi^l(k)}^l, \bar{y}^l\right) \leq \frac{R_x}{2^l} \qquad \forall k \geq k_l.$$

Define the sequence $\{z_l\}_l$ by

$$z_l := y_{(\varphi^1 \circ \varphi^2 \circ \cdots \circ \varphi^l)(k_l)} \qquad \forall l.$$

Then proceeding as above shows that for every $l \geq 1$, one has

$$d_{SR}(z_l, z_{l+1}) \leq \frac{4R_x}{2^l}.$$

Hence $\{z_k\}_k$ is a Cauchy sequence in $\bar{B}_{SR}(x, R_x)$. Since (M, d_{SR}) is complete, it converges to some $z \in \bar{B}_{SR}(x, R_x)$.   □

Lemma 2.6 *The interval I_x is open in $[0, +\infty)$.*

Proof (Proof of Lemma 2.6) We need to show that if $R \in I_x$, then there is $\delta > 0$ such that $R + \delta$ belongs to I_x. Let $R > 0$ in I_x be fixed. Denote by $\partial B_{SR}(x, R)$ the boundary of $\bar{B}_{SR}(x, R)$, that is $\partial B_{SR}(x, R) = \bar{B}_{SR}(x, R) \setminus B_{SR}(x, R)$. Since $\bar{B}_{SR}(x, R)$ is assumed to be compact, its boundary is compact too. From Proposition 2.3, we know that for every $y \in \partial B_{SR}(x, R)$, there is $\delta_y > 0$ such that $\bar{B}_{SR}(y, 2\delta_y)$ is compact. Since

$$\partial B_{SR}(x, R) \subset \cup_{y \in \partial B_{SR}(x,R)} B_{SR}\left(y, \delta_y\right),$$

there is a finite number of points $y_1, \ldots, y_N$ in $\partial B_{SR}(x, R)$ such that

$$\partial B_{SR}(x, R) \subset \cup_{i=1}^{N} B_{SR}\left(y_i, \delta_{y_i}\right).$$

Set

$$\delta = \min \left\{ \frac{\delta_{y_i}}{2} \mid i = 1, \ldots, N \right\}.$$

We prove easily that

$$\bar{B}_{SR}(x, R + \delta) \subset \left(\bar{B}_{SR}(x, R) \cup \cup_{i=1}^{N} \bar{B}_{SR}\left(y_i, 2\delta_{y_i}\right) \right)$$

which is a finite union of compact sets, hence compact as well. This shows that $\bar{B}_{SR}(x, R + \delta)$ is compact.   □

In conclusion, I_x is both open and closed in $[0, +\infty)$. Hence $I_x = [0, +\infty)$ which concludes the proof of (i). Let us now prove assertion (ii). We note that since Δ does not necessarily admit a global orthonormal frame on M, we cannot repeat verbatim the proof of Proposition 2.3. Let $x, y \in M$ be fixed, set $R := \max\{2d_{SR}(x, y), 1\}$. By (i), we know that $\bar{B}_{SR}(x, R)$ is compact. Let $\{\gamma^k\}_k$ be a sequence of horizontal paths with constant speed in $\Omega_{\Delta}^{x,1}$ joining x to y such that

$$d_{SR}(x, y) = \lim_{k \to +\infty} \text{length}(\gamma^k).$$

Without loss of generality we may assume that

$$\text{length}(\gamma^k) < R \qquad \forall k,$$

which means that all the curves γ^k remain in $\bar{B}_{SR}(x, R)$. By Proposition 2.3, for every $z \in \bar{B}_{SR}(x, R)$ there is $\rho_z > 0$ such that any minimizing sequence of horizontal paths with constant speed contained in $B_{SR}(z, \rho_z)$ converges uniformly (up to taking a subsequence) to some minimizing geodesic. By compactness, there are $z_1, \ldots, z_L \in \bar{B}_{SR}(x, R)$ and an integer $N > 1$ with $R/N < \min\{\rho_1, \ldots, \rho_L\}/4$ such that

$$B_{SR}(x, R) \subset \bigcup_{l=1}^{L} B_{SR}(z_l, 1/N).$$

Set for every $j = 0, \ldots, N, t_j := j/N$, for every $j = 0, \ldots, N-1, I_j := [t_j, t_{j+1}]$, and denote by γ_j^k the restriction of γ^k to the interval I_j. Fix $j \in \{0, \ldots, N-1\}$. For every k, there is $l \in \{1, \ldots, L\}$ (which may depend on k) such that $d_{SR}(\gamma^k(t_j), z_l) < 1/N$, then

$$d_{SR}\left(\gamma^k(t), z_l\right) \leq d_{SR}(\gamma^k(t_j), z_l) + \frac{\text{length}^g(\gamma^k)}{N} < \frac{1}{N} + \frac{R}{N} < \rho_l,$$

for every $t \in I_j$. This shows that each piece of horizontal path γ_j^k with length $\text{length}^g(\gamma^k)/N$ is contained in some $B_{SR}(z_l, \rho_l)$. Therefore, up to taking a subsequence, the sequence $\{\gamma_j^k\}_k$ converges to some minimizing geodesic with length $d_{SR}(x, y)/N$. We deduce easily the existence of a subsequence of $\{\gamma^k\}_k$ converging to some minimizing geodesic between x and y. $\square$

Remark 2.2 In fact, we proved a global version of Proposition 2.3. If (M, d_{SR}) is a complete metric space, then for every $x, y \in M$ and every minimizing sequence $\{\gamma^k\}_k$ of horizontal paths with constant speed in $\Omega_\Delta^{x,1}$ joining x to y such that

$$d_{SR}(x, y) = \lim_{k \to +\infty} \text{length}(\gamma^k),$$

up to taking a subsequence, $\{\gamma^k\}_k$ converges uniformly to some minimizing geodesic joining x to y.

We shall say that the sub-Riemannian structure (Δ, g) on M is *complete* if the metric space (M, d_{SR}) is complete. The following result holds.

Proposition 2.7 *Let (Δ, g) be a sub-Riemannian structure on M, assume that (M, g) is a complete Riemannian manifold. Then for any totally nonholonomic distribution Δ, the SR structure (Δ, g) on M is complete.*

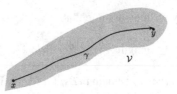

Fig. 2.1 An orthonormal frame along γ

Proof Denote by d_g the Riemannian geodesic distance on M with respect to g. Since the set of paths joining x to y contains the set of horizontal paths joining x to y, there holds

$$d_g(x, y) \leq d_{SR}(x, y) \qquad \forall x, y \in M.$$

Therefore, any Cauchy sequence with respect to d_{SR} is a Cauchy sequence with respect to d_g. Hence it is convergent. Since both topology coincide, it is convergent with respect to d_{SR} as well.   □

2.2 The Hamiltonian Geodesic Equation

Throughout all the section, we assume that the SR structure (Δ, g) is complete. Thanks to Theorem 2.4, minimizing geodesics exist between any pair of points in M.

Normal and abnormal geodesics. Let $x, y \in M$ and a minimizing geodesic $\gamma \in \Omega_\Delta^{x,1}$ joining x to y be fixed. Since γ minimizes the distance between x and y it cannot have self-intersection. Hence (Δ, g) admits an orthonormal frame along γ (Fig. 2.1).

Then there is an open neighborhood $\mathcal{V}$ of $\gamma([0, 1])$ in M and an orthonormal family $\mathcal{F}$ (with respect to the metric g) of m smooth vector fields $X^1, \ldots, X^m$ such that

$$\Delta(z) = \mathrm{Span}\left\{X^1(z), \ldots, X^m(z)\right\} \qquad \forall z \in \mathcal{V}.$$

Moreover, there is a control $u^\gamma \in L^2([0, 1]; \mathbb{R}^m)$ (which indeed belong to the open set $\mathcal{U}_{\mathcal{F}}^{x,1}$ which was defined in Proposition 1.7) such that

$$\dot{\gamma}(t) = \sum_{i=1}^m u_i^\gamma(t) X^i(\gamma(t)) \qquad \text{a.e. } t \in [0, 1].$$

Since γ is a minimizing geodesic between x and y, it minimizes the energy among all horizontal paths joining x to y. Since there is a local one-to-one correspondence between the set of horizontal paths starting at x and the set of trajectories of some control system (see Proposition 1.7), the control u^γ minimizes the quantity

$$\int_0^1 g_{\gamma_u(t)}\left(\sum_{i=1}^m u_i(t)X^i(\gamma_{x,u}(t)), \sum_{i=1}^m u_i(t)X^i(\gamma_u(t))\right)dt = \int_0^1 \sum_{i=1}^m u_i(t)^2 dt =: C(u),$$

among all controls $u \in L^2([0, 1]; \mathbb{R}^m)$ such that the solution $\gamma_u : [0, 1] \to M$ of the Cauchy problem

$$\dot{\gamma}_u(t) = \sum_{i=1}^k u_i(t)X^i(\gamma_u(t)) \quad \text{a.e. } t \in [0, 1], \qquad \gamma_u(0) = x,$$

is well-defined on $[0, 1]$ and satisfies (the End-Point mapping $E_{\mathscr{F}}^{x,1}$ has been defined in Chap. 1)

$$E_{\mathscr{F}}^{x,1}(u) = y.$$

In other terms, there is an open set $\mathscr{U} \subset L^2([0, 1]; \mathbb{R}^m)$ such that u^γ is solution to the following optimization problem:

$$u^\gamma \text{ minimizes } C(u) \text{ among all } u \in \mathscr{U} \text{ with } E_{\mathscr{F}}^{x,1}(u) = 1.$$

By the Lagranges Multipliers Theorem (see Theorem B.2), there is $p \in T_y^*M \simeq (\mathbb{R}^n)^*$ and $\lambda_0 \in \{0, 1\}$ with $(\lambda_0, p) \neq (0, 0)$ such that

$$p \cdot D_{u^\gamma}E_{\mathscr{F}}^{x,1}(v) = \lambda_0 D_{u^\gamma}C(v) \qquad \forall v \in L^2([0, 1]; \mathbb{R}^m). \tag{2.3}$$

Two cases may appear, either $\lambda_0 = 0$ or $\lambda_0 = 1$. By restricting $\mathscr{V}$ if necessary, we can assume that the cotangent bundle T^*M is trivializable with coordinates $(x, p) \in \mathscr{V} \times (\mathbb{R}^n)^*$ over $\mathscr{V}$.

First case: $\lambda_0 = 0$.
Then we have $p \in T_y^*M \setminus \{0\} \simeq (\mathbb{R}^n)^* \setminus \{0\}$ satisfying

$$p \cdot D_{u^\gamma}E_{\mathscr{F}}^{x,1}(v) = 0 \qquad \forall v \in L^2([0, 1]; \mathbb{R}^m).$$

This means that some nonzero linear form annihilates the image of $E_{\mathscr{F}}^{x,1}$. Then u^γ is singular with respect to x and $\mathscr{F}$ or equivalently the path γ is singular with respect to Δ. By Proposition 1.11 and Remark 1.8, γ admits an abnormal extremal lift, that is there is an absolutely continuous arc $p : [0, 1] \to (\mathbb{R}^n)^* \setminus \{0\}$ with $p(1) = p$ which satisfies

$$\dot{p}(t) = -\sum_{i=1}^k u_i(t)\,p(t) \cdot D_{\gamma(t)}X^i \qquad \text{a.e. } t \in [0, 1]$$

and

$$p(t) \cdot X^i(\gamma(t)) = 0, \quad \forall t \in [0, 1] \quad \forall i = 1, \cdots, m.$$

In other terms, γ is a *singular minimizing geodesic*.

Second case: $\lambda_0 = 1$.

  Define in local coordinates, the Hamiltonian $H : \mathscr{V} \times (\mathbb{R}^n)^* \to \mathbb{R}$ by

$$H(x, p) := \frac{1}{2} \sum_{i=1}^m (p \cdot X^i(x))^2 = \max_{u \in \mathbb{R}^m} \left\{ \sum_{i=1}^m u_i \, p \cdot X^i(x) - \frac{1}{2} \sum_{i=1}^m u_i^2 \right\} \quad (2.4)$$

for all $(x, p) \in \mathscr{V} \times (\mathbb{R}^n)^*$. Then the following result holds.

Proposition 2.8 *Equality (2.3) with $\lambda_0 = 1$ yields the existence of a smooth arc $p : [0, 1] \longrightarrow (\mathbb{R}^n)^*$ with $p(1) = \frac{p}{2}$, such that the pair (γ, p) satisfies*

$$\begin{cases} \dot{\gamma}(t) = \frac{\partial H}{\partial p}(\gamma(t), p(t)) = \sum_{i=1}^m \left[p(t) \cdot X^i(\gamma(t)) \right] X^i(\gamma(t)) \\ \dot{p}(t) = -\frac{\partial H}{\partial x}(\gamma(t), p(t)) = -\sum_{i=1}^m \left[p(t) \cdot X^i(\gamma(t)) \right] p(t) \cdot D_{\gamma(t)} X^i \end{cases} \quad (2.5)$$

for a.e. $t \in [0, 1]$ and

$$u_i^\gamma(t) = p(t) \cdot X^i(\gamma(t)) \quad \text{for a.e. } t \in [0, 1], \quad \forall i = 1, \ldots, m. \quad (2.6)$$

In particular, the path γ is smooth on $[0, 1]$.

Proof The differential of $C : L^2([0, 1]; \mathbb{R}^m) \to \mathbb{R}$ at u^γ is given by

$$D_{u^\gamma} C(v) = 2 \langle u^\gamma, v \rangle_{L^2} \quad \forall v \in L^2([0, 1]; \mathbb{R}^m).$$

Moreover by Remark 1.5, the differential of $E_{\mathscr{F}}^{x,1}$ at u^γ is given by

$$D_{u^\gamma} E_{\mathscr{F}}^{x,1}(v) = S(1) \int_0^1 S(t)^{-1} B(t) v(t) dt \quad \forall v \in L^2([0, 1]; \mathbb{R}^m),$$

where the functions A, B, S were defined in Remark 1.5. Hence (2.3) yields

$$\int_0^1 \left[p \cdot S(1) S(t)^{-1} B(t) - 2 u^\gamma(t)^* \right] v(t) dt = 0 \quad \forall v \in L^2([0, 1]; \mathbb{R}^m).$$

Which implies

$$u^\gamma(t) = \frac{1}{2} \left(p \cdot S(1) S(t)^{-1} B(t) \right)^* \quad \text{a.e. } t \in [0, 1].$$

Let us define $p : [0, 1] \to (\mathbb{R}^n)^*$ by

$$p(t) := \frac{1}{2} p \cdot S(1) S(t)^{-1} \qquad \forall t \in [0, 1].$$

By construction, for a.e. $t \in [0, 1]$ we have $u^\gamma(t)^* = p(t) \cdot B(t)$, which means that (2.6) is satisfied. Furthermore, as in the proof of Proposition 1.11, we have $\dot{p}(t) = -p(t) \cdot A(t)$ for a.e. $t \in [0, 1]$. This means that (2.5) is satisfied for a.e. $t \in [0, 1]$. The pair (γ, p) is solution to a smooth autonomous differential equation, hence it is smooth. $\qquad\square$

The curve $\psi : [0, 1] \to T^*M$ given by $\psi(t) = (\gamma(t), p(t))$ for every $t \in [0, 1]$ is a *normal extremal* whose projection is γ and which satisfies $\psi(1) = (y, \frac{p}{2})$. We say that ψ is a *normal extremal lift* of γ. We also say that γ is a *normal minimizing geodesic*.

Define the *sub-Riemannian Hamiltonian* $H : T^*M \to \mathbb{R}$ as follows. For every $x \in M$, the restriction of H to the fiber T_x^*M is given by the nonnegative quadratic form

$$p \longmapsto \frac{1}{2} \max \left\{ \frac{p(v)^2}{g_x(v, v)} \mid v \in \Delta(x) \setminus \{0\} \right\}. \qquad (2.7)$$

Let $\overrightarrow{H}$ denote the Hamiltonian vector field on T^*M associated to H, that is, $\iota_{\overrightarrow{H}} \omega = -dH$, or in local coordinates

$$\overrightarrow{H}(x, p) = \left(\frac{\partial H}{\partial p}(x, p), -\frac{\partial H}{\partial x}(x, p) \right).$$

A *normal extremal* is an integral curve of $\overrightarrow{H}$ defined on some interval $[0, T]$, i.e., a curve $\psi : [0, T] \to T^*M$ such that $\dot{\psi}(t) = \overrightarrow{H}(\psi(t))$, for $t \in [0, T]$. The projection of a normal extremal $\psi : [0, T] \to T^*M$ is a smooth horizontal path $\gamma := \pi \circ \psi : [0, T] \to M$ with constant speed given by

$$\left| \dot{\gamma}(t) \right|_{\gamma(t)}^g = \sqrt{2H(\psi(t))} \qquad \forall t \in [0, T].$$

We check easily that the Hamiltonian defined by (2.7) reads as (2.4) in local coordinates. Then the previous study yields the following result.

Theorem 2.9 *Let $\gamma : [0, 1] \to M$ be a minimizing geodesic between x and y in M. One of the two following non-exclusive cases occur:*

- *γ is singular.*
- *γ admits a normal extremal lift in T^*M.*

Be careful, a minimizing geodesic could be both singular and the projection of a normal extremal. In Sect. 2.5, we shall see several examples of minimizing geodesics, including the cases of singular normal minimizing geodesics and strictly abnormal minimizing geodesic, that is abnormal geodesics admitting no normal extremal lift.

Remark 2.3 In the Riemannian case, that is if Δ has rank $m = n$, any path is horizontal and regular (see Remark 1.9). As a consequence any minimizing geodesic is normal.

Short normal geodesics are minimizing. Projections of normal extremals are minimizing for short times.

Proposition 2.10 *Let $\bar{x} \in M$ and $\bar{p} \in T_{\bar{x}}^* M$ with $H(\bar{x}, \bar{p}) \neq 0$ be fixed. Then there is a neighborhood $\mathscr{W}$ of $\bar{p}$ in $T_{\bar{x}}^* M$ and $\varepsilon > 0$ such that every normal extremal so that $\psi(0) = (\bar{x}, p)$ (in local coordinates) belongs to $\mathscr{W}$ minimizes the SR energy on the interval $[0, \varepsilon]$. That is if we set $\gamma := \pi \circ \psi : [0, \varepsilon] \to M$, then we have*

$$e_{SR}(\gamma(0), \gamma(\varepsilon)) = 2H(x, p)\varepsilon^2.$$

In particular, γ minimizes the length between $\bar{x}$ and $\gamma(\varepsilon)$.

Proof Since the result is local, we can assume that we work in $\mathbb{R}^n$. Then we can assume that (Δ, g) admits an orthonormal frame $\mathscr{F} = \{X^1, \ldots, X^m\}$. For sake of simplicity, we identify $(\mathbb{R}^n)^*$ with $\mathbb{R}^n$. Then the Hamiltonian $H : \mathbb{R}^n \times \mathbb{R}^n \to \mathbb{R}$ which were defined in (2.4) and (2.7) is given by

$$H(x, p) := \max_{u \in \mathbb{R}^m} \left\{ \langle p, \sum_{i=1}^m u_i X^i(x) \rangle - \frac{1}{2} \sum_{i=1}^m u_i^2 \right\} = \frac{1}{2} \sum_{i=1}^m \langle p, X^i(x) \rangle^2,$$

for every $(x, p) \in R^n \times \mathbb{R}^n$.

Our aim is now to prove the following result: for every $p_0 \in \mathbb{R}^n$ such that $H(\bar{x}, p_0) \neq 0$, there exist a neighborhood $\mathscr{W}$ of p_0 in $\mathbb{R}^n$ and $\varepsilon > 0$ such that every solution $(x, p) : [0, \varepsilon] \to \mathbb{R}^n \times \mathbb{R}^n$ of the Hamiltonian system

$$\begin{cases} \dot{x}(t) = \dfrac{\partial H}{\partial p}(x(t), p(t)) = \displaystyle\sum_{i=1}^m \langle p(t), X^i(x(t)) \rangle X^i(x(t)) \\[2mm] \dot{p}(t) = -\dfrac{\partial H}{\partial x}(x(t), p(t)) = -\displaystyle\sum_{i=1}^m \langle p(t), X^i(x(t)) \rangle \left(D_{x(t)} X^i\right)^*(p(t)), \end{cases} \quad (2.8)$$

with $x(0) = \bar{x}$ and $p(0) \in \mathscr{W}$, satisfies

$$2\varepsilon H(\bar{x}, p_0) = \int_0^\varepsilon \sum_{i=1}^m \langle p(t), X^i(x(t)) \rangle^2 dt \leq \int_0^\varepsilon \sum_{i=1}^m u_i(t)^2 dt, \quad (2.9)$$

for every control $u \in L^2([0, \varepsilon]; \mathbb{R}^m)$ such that the solution of

$$\dot{y}(t) = \sum_{i=1}^{m} u_i(t) X^i(y(t)), \quad y(0) = \bar{x}, \tag{2.10}$$

satisfies $y(\varepsilon) = x(\varepsilon)$. Let $p_0 \in \mathbb{R}^n$ with $H(\bar{x}, p_0) \neq 0$ be fixed, we need the following lemma.

Lemma 2.11 *There exist a neighborhood $\mathcal{W}$ of p_0 and $\rho > 0$ such that, for every $p \in \mathcal{W}$, there exists a function $S : B(\bar{x}, \rho) \to \mathbb{R}$ of class C^1 which satisfies*

$$H(x, \nabla S(x)) = H(\bar{x}, p), \quad \forall x \in B(\bar{x}, \rho), \tag{2.11}$$

and such that, if $(x^p, p^p) : [-\rho, \rho] \to \mathbb{R}^n \times \mathbb{R}^n$ denotes the solution of (2.8) satisfying $x^p(0) = \bar{x}$ and $p^p(0) = p$, then

$$\nabla S(x^p(t)) = p^p(t), \quad \forall t \in (-\rho, \rho). \tag{2.12}$$

Proof (Proof of Lemma 2.11) The proof consists in applying the *method of characteristics*. Let Π be the linear hyperplane such that $\langle p_0, v \rangle = 0$ for every $v \in \Pi$. We first show how to construct locally S as the solution of the Hamilton-Jacobi Equation (2.11) which vanishes on $\bar{x} + \Pi$ and such that $\nabla S(\bar{x}) = p_0$. Up to considering a smaller neighborhood $\mathcal{V}$, we assume that $H(x, p_0) \neq 0$ for every $x \in \mathcal{V}'$. For every $x \in (\bar{x} + \Pi) \cap \mathcal{V}$, set

$$\bar{p}(x) := \sqrt{\frac{H(\bar{x}, p_0)}{H(x, p_0)}} p_0.$$

Then, $H(x, \bar{p}(x)) = H(\bar{x}, p_0)$ and $\bar{p}(x) \perp \Pi$, for every $x \in \mathcal{V}'$. There exists $\mu > 0$ such that, for every $x \in (\bar{x} + \Pi) \cap \mathcal{V}$, the solution (x_x, p_x) of (2.8), satisfying $x_x(0) = x$ and $p_x(0) = \bar{p}(x)$, is defined on the interval $(-\mu, \mu)$.

For every $x \in (\bar{x} + \Pi) \cap \mathcal{V}$ and every $t \in (-\mu, \mu)$, set $\theta(t, x) := x_x(t)$. The mapping $(t, x) \mapsto \theta(t, x)$ is smooth. Moreover, $\theta(0, x) = x$ for every $x \in (\bar{x} + \Pi) \cup \mathcal{V}$ and $\dot{\theta}(0, \bar{x}) = \sum_{i=1}^{m} \langle \bar{p}(x), X^i(\bar{x}) \rangle X^i(\bar{x})$ does not belong to Π. Hence there exists $\rho \in (0, \mu)$ with $B(\bar{x}, \rho) \subset \mathcal{V}$ such that the mapping θ is a smooth diffeomorphism from $(-\rho, \rho) \times ((\bar{x} + \Pi) \cap B(\bar{x}, \rho))$ into a neighborhood $\mathcal{V}'$ of $\bar{x}$. Denote by $\varphi = (\tau, \pi)$ the inverse function of θ, that is the function such that $(\theta \circ \varphi)(x) = (\tau(x), \pi(x)) = x$ for every $x \in \mathcal{V}'$. Define the two vector fields X and P by

$$X(x) := \dot{\theta}(\tau(x), \pi(x)) \quad \text{and} \quad P(x) := p_{\pi(x)}(\tau(x)), \quad \forall x \in \mathcal{V}'.$$

Then,

$$X(\theta(t,x)) = \dot{\theta}(t,x) = \dot{x}_x(t) = \sum_{i=1}^{m} \langle p_x(t), X^i(x_x(t)) \rangle X^i(x_x(t))$$

$$= \sum_{i=1}^{m} \langle P(\theta(t,x)), X^i(\theta(t,x)) \rangle X^i(\theta(t,x)),$$

and

$$\sum_{i=1}^{m} \langle P(\theta(t,x)), X^i(x_x(t)) \rangle^2 = \sum_{i=1}^{m} \langle p_x(t), X^i(x_x(t)) \rangle^2 = 2H(x, \bar{p}(x)) = 2H(\bar{x}, p_0),$$

for every $t \in (-\rho, \rho)$ and every $x \in (\bar{x} + \Pi) \cap B(\bar{x}, \rho)$. For every $x \in \mathscr{V}'$, set $\alpha_i(x) := \langle P(x), X^i(x) \rangle$. Hence,

$$X(x) = \sum_{i=1}^{m} \alpha_i(x) X^i(x) \quad \text{and} \quad \sum_{i=1}^{m} \alpha_i(x)^2 = 2H(\bar{x}, p_0),$$

for every $x \in \mathscr{V}'$. Define the function $S : \mathscr{V}' \mapsto \mathbb{R}$ by

$$S(x) := 2H(\bar{x}, p_0)\tau(x), \quad \forall x \in \mathscr{V}'.$$

We next prove that $\nabla S(x) = P(x)$ for every $x \in \mathscr{V}'$. For every $t \in (-\rho, \rho)$, denote by $W_t := \{y \in \mathscr{V}' \mid \tau(y) = t\}$. In fact, W_t coincides with the set of $y \in \mathscr{V}'$ such that $S(y) = 2H(\bar{x}, p_0)t$. It is a smooth hypersurface which satisfies $\nabla S(y) \perp T_y W_t$ for every $y \in W_t$. Let $y \in W_t$ be fixed, there exists $x \in (\bar{x} + \Pi) \cup B(\bar{x}, \rho)$ such that $y = \theta(t, x) = x_x(t)$. Let us first prove that $P(y) = p_x(t)$ is orthogonal to $T_y W_t$. To this aim, without loss of generality we assume that $t > 0$. Let $w \in T_y W_t$, there exists $v \in \Pi$ such that $w = D_x \theta_t(v)$. For every $s \in [0, t]$, set $z(s) := D_x \theta(s, x)(v)$. We have

$$\dot{z}(s) = \frac{d}{ds} D_x \theta(s, x)v = \frac{d}{dx} \dot{\theta}(t, x)v = \frac{d}{dx} X(\theta(t,x))v = D_{\theta(t,x)} X(z(s)).$$

Hence,

$$\frac{d}{ds} \langle z(s), p_x(s) \rangle = \langle \dot{z}(s), p_x(s) \rangle + \langle z(s), \dot{p}_x(s) \rangle$$

$$= \langle D_{\theta(s,x)} X z(s), p_x(s) \rangle$$

$$- \langle z(s), \sum_{i=1}^{m} \langle p_x(s), X^i(x_x(s)) \rangle \left(D_{x_x(s)} X^i \right)^* \left(p_x(s) \right) \rangle.$$

Since $X(x) = \sum_{i=1}^{m} \alpha_i(x) X^i(x)$ and $\sum_{i=1}^{m} \alpha_i(x)^2 = 2H(\bar{x}, p_0)$ for every $x \in \mathscr{V}'$, there holds

$$\left(D_{x_x(s)}X\right)^*\left(p_x(s)\right) = \sum_{i=1}^{m} \alpha_i(x_x(s))\left(D_{x_x(s)}X^i\right)^*\left(p_x(s)\right) + \sum_{i=1}^{m} \langle X^i(x_x(s)), p_x(s)\rangle \nabla\alpha_i(x_x(s))$$

$$= \sum_{i=1}^{m} \alpha_i(x_x(s))\left(D_{x_x(s)}X^i\right)^*\left(p_x(s)\right) + \sum_{i=1}^{m} \alpha_i(x_x(s))\nabla\alpha_i(x_x(s))$$

$$= \sum_{i=1}^{m} \alpha_i(x_x(s))\left(D_{x_x(s)}X^i\right)^*\left(p_x(s)\right).$$

We deduce that $\frac{d}{ds}\langle z(s), p_x(s)\rangle = 0$ for every $s \in [0, t]$. Hence,

$$\langle w, P(y)\rangle = \langle w, p_x(t)\rangle = \langle z(t), p_x(t)\rangle = \langle z(0), \bar{p}(x)\rangle = 0.$$

This proves that $P(y)$ is orthogonal to $T_y W_t$, which implies that $P(y)$ and $\nabla S(y)$ are colinear. Furthermore, since $S(x_x(s)) = 2H(\bar{x}, p_0)s$ for every $s \in [0, t]$, one gets

$$\langle \nabla S(x_x(t)), \dot{x}_x(t)\rangle = 2H(\bar{x}, p_0) = \langle p_x(t), \dot{x}_x(t)\rangle.$$

Since $\dot{x}_x(t) = X(y)$ does not belong to $T_y W_t$, we deduce that $\nabla S(x_x(t)) = p_x(t)$. In consequence, we proved that $\nabla S(x) = P(x)$ for every $x \in \mathcal{V}'$. □

Let us now conclude the proof of Proposition 2.10. Clearly, there exists $\varepsilon > 0$ such that every solution $(x, p) : [0, \varepsilon] \to \mathbb{R}^n \times \mathbb{R}^n$ of (2.8), with $x(0) = \bar{x}$ and $p(0) \in \mathcal{W}$, satisfies

$$x(t) \in B(\bar{x}, \rho), \quad \forall t \in [0, \varepsilon].$$

Moreover, we have by (2.11)–(2.12)

$$S(x(\varepsilon)) - S(\bar{x}) = 2\varepsilon H(\bar{x}, p).$$

Let $u \in L^2([0, \varepsilon]; \mathbb{R}^m)$ be a control such that the solution $y : [0, \varepsilon] \to \mathcal{W}$ of (2.10) starting at $\bar{x}$ satisfies $y(\varepsilon) = x(\varepsilon)$. We have

$$S(x(\varepsilon)) - S(\bar{x}) = S(y(\varepsilon)) - S(y(0))$$

$$= \int_0^\varepsilon \frac{d}{dt}\left(S(y(t))\right) dt$$

$$= \int_0^\varepsilon \langle \nabla S(y(t)), \dot{y}(t)\rangle dt$$

$$\leq \int_0^\varepsilon H(y(t), \nabla S(y(t))) + \frac{1}{2}\sum_{i=1}^{m} u_i(t)^2 dt$$

$$= \varepsilon H(\bar{x}, p) + \frac{1}{2}\int_0^\varepsilon \sum_{i=1}^{m} u_i(t)^2 dt.$$

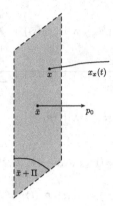

Fig. 2.2 The method of characteristics

Inequality (2.9) follows.   □

2.3 The Sub-Riemannian Exponential Map

Definition. Recall that the SR Hamiltonian $H : T^*M \to \mathbb{R}$ which is canonically associated with our SR structure (Δ, g) is defined by

$$H(x, p) = \frac{1}{2} \max \left\{ \frac{p(v)^2}{g_x(v, v)} \mid v \in \Delta(x) \setminus \{0\} \right\} \qquad \forall (x, p) \in T^*M.$$

We recall that a normal extremal is a curve $\psi : [0, T] \to T^*M$ satisfying

$$\dot{\psi}(t) = \vec{H}(\psi(t)) \qquad \forall t \in [0, T].$$

Let $x \in M$ be fixed. We first define the domain $\mathscr{E}_x \subset T_x^*M$ of the SR exponential map by,

$$\mathscr{E}_x := \left\{ p \in T_x^*M \mid \psi_{x,p} \text{ is defined on the interval } [0, 1] \right\},$$

where $\psi_{x,p}$ is the normal extremal so that $\psi_{x,p}(0) = (x, p)$ in local coordinates. The set $\mathscr{E}_x$ is an open subset of T_x^*M containing the origin and star-shaped with respect to 0.

The *sub-Riemannian exponential map* from x is defined by

$$\exp_x : \mathscr{E}_x \subset T_x^*M \longrightarrow M$$
$$p \longmapsto \pi\left(\psi_{x,p}(1)\right).$$

By rescaling, if $(x_p, p_p) : [0, T] \to T^*M$ is the trajectory of the Hamiltonian vector field $\vec{H}$ with $x(0) = x, p(0) = p$, then we have

$$\big(x_p(\lambda t), \lambda p_p(\lambda t)\big) = \big(x_{\lambda p}(t), p_{\lambda p}(t)\big) \quad \forall t \in [0, T/\lambda], \ \forall \lambda > 0.$$

Then, for every $p \in T_x^*M$, the curve $\gamma_p : [0, 1] \to M$ defined by

$$\gamma_p(t) := \exp_x(tp) = \pi\big(\psi_{x,p}(t)\big) \quad \forall t \in [0, 1],$$

is an horizontal path with constant speed satisfying

$$\text{energy}^g\,(\pi(\psi_{x,p})) = \big(\text{length}^g\,(\pi(\psi_{x,p}))\big)^2 = 2H\big(\psi_{x,p}(0)\big) = 2H(x, p).$$

Proposition 2.13 *Assume that* (Δ, g) *is complete. Then*

$$\mathscr{E}_x = T_x^*M \quad \forall x \in M.$$

Proof We argue by contradiction. Let $\bar{x} \in M$ and $\psi = (\bar{\gamma}, p_{\bar{p}}) : [0, T) \to T^*M$ be a normal extremal starting at $(\bar{x}, \bar{p}) \in T_{\bar{x}}^*M$ that extends to no interval $[0, T + \varepsilon)$ for $\varepsilon > 0$. Let $\{t_k\}_k$ be any increasing sequence that approaches T, and set $y_k := \bar{\gamma}(t_k)$. Since $\bar{\gamma}$ is an horizontal path with constant speed $V = \sqrt{2H(\bar{x}, \bar{p})}$, we have

$$d_{SR}(y_k, y_l) \le V |t_k - t_l| \quad \forall k, l.$$

Then $\{y_k\}_k$ is a Cauchy sequence in M. By completeness $\{y_k\}_k$ converges to some point $y \in M$. Let $\{X^1, \dots, X^m\}$ be a local orthonormal frame in a small ball $B_{SR}(y, r)$. In local coordinates near y, H reads

$$H(x, p) = \frac{1}{2} \sum_{i=1}^{m} \big(p \cdot X^i(x)\big)^2$$

and $(\bar{\gamma}, p_{\bar{p}})$ satisfies the differential system

$$\begin{cases} \dot{\bar{\gamma}}(t) = \frac{\partial H}{\partial p}\big(\bar{\gamma}(t), p_{\bar{p}}(t)\big) = \sum_{i=1}^{m} \big[p_{\bar{p}}(t) \cdot X^i\big(\bar{\gamma}(t)\big)\big] X^i\big(\bar{\gamma}(t)\big) \\ \dot{p}_{\bar{p}}(t) = -\frac{\partial H}{\partial x}\big(\bar{\gamma}(t), p_{\bar{p}}(t)\big) = -\sum_{i=1}^{m} \big[p_{\bar{p}}(t) \cdot X^i\big(\bar{\gamma}(t)\big)\big] p_{\bar{p}}(t) \cdot D_{\bar{\gamma}(t)}X^i, \end{cases}$$

for $t \in [T - \delta, T)$ with $\delta > 0$ small enough. Since H is constant along $(\bar{\gamma}, p_{\bar{p}})$, we have

$$\big|p_{\bar{p}}(t) \cdot X^i\big(\bar{\gamma}(t)\big)\big| \le \sqrt{m}V \quad \forall i = 1, \dots, m,$$

and by compactness, the vector fields $X_1, \ldots, X_m$ and their differentials are bounded in $\bar{B}_{SR}(y, r)$. Thus there is a constant $K > 0$ such that

$$\left| \dot{p}_{\bar{p}}(t) \right| \leq K \left| p_{\bar{p}}(t) \right| \qquad \forall t \in [T - \delta, T).$$

By Gronwall's Lemma (see Lemma A.1), we infer that both $\bar{\gamma}$ and $p_{\bar{p}}$ are uniformly bounded near T. This means that the extremal ψ can be extended beyong T, which gives a contradiction. $\qquad\qquad\qquad\qquad\qquad\qquad\qquad\qquad\qquad\qquad\qquad\qquad\square$

Remark 2.4 Let $(x, p) \in T^*M$ such that γ_p is singular be fixed. By Proposition 1.11, γ_p is the projection of an abnormal extremal $\psi : [0, 1] \rightarrow T^*M$ (written as (γ_p, q) in local coordinates). Then taking local coordinates, for every $\lambda \in \mathbb{R}$ the curve (here $\psi_{x,p} = (\gamma_p, p_p)$ denotes the normal extremal starting at (x, p))

$$t \in [0, 1] \longmapsto \psi_{x,p}(t) + \lambda \psi(t) = \left(\gamma_p(t), p_p(t) + \lambda q(t) \right)$$

is a normal extremal starting at $(x, p + \lambda q)$. Then we have

$$\exp_x \left(p + \lambda q \right) = \exp_x(p) \qquad \forall \lambda \in \mathbb{R}.$$

Remark 2.5 Let (Δ, g) be a complete sub-Riemannian structure on M and $x \in M$ be fixed. If (Δ, g) does not admit singular minimizing curves from x, then the exponential map from x is onto. As a matter of fact, for every $y \in M$, there is a minimizing geodesic $\gamma : [0, 1] \rightarrow M$ joining x to y. Since γ is not singular, it is the projection of a normal extremal (see Theorem 2.9), which means that there is $p \in T_x^*M$ such that $\exp_x(p) = y$.

On the image of the Sub-Riemannian exponential map. The functions $\exp_x$ are "almost" onto.

Theorem 2.14 *Assume that (Δ, g) is complete and let $x \in M$ be fixed. There is an open and dense set $\mathscr{D} \subset M$ such that for every $y \in \mathscr{D}$ there is $p_y \in T_x^*M$ satisfying*

$$\exp_x(p_y) = y \quad \text{and} \quad d_{SR}(x, y) = \sqrt{2H\left(x, p_y\right)}.$$

In particular, the set $\exp_x(T_x^*M)$ contains an open dense subset of M.

Proof Let us begin with a preparatory lemma.

Lemma 2.15 *Let $y \neq x$ in M be such that there is a function $\phi : M \rightarrow \mathbb{R}$ differentiable at y such that*

$$\phi(y) = d_{SR}^2(x, y) \quad \text{and} \quad d_{SR}^2(x, z) \geq \phi(z) \ \forall z \in M.$$

Then there is a unique minimizing geodesic $\gamma : [0, 1] \rightarrow M$ between x and y. It is the projection of a normal extremal $\psi : [0, 1] \rightarrow T^*M$ satisfying $\psi(1) = (y, \frac{1}{2}D_y\phi)$. In particular $x = \exp_y(-\frac{1}{2}D_y\phi)$.

Proof (Proof of Lemma 2.15) Since $e_{SR}(x, z) = d_{SR}^2(x, z)$ for any $z \in M$, the assumption of the proposition implies that there is a neighborhood $\mathscr{U}$ of y in M such that

$$e_{SR}(x, z) \geq \phi(z) \quad \forall z \in \mathscr{U} \quad \text{and} \quad e_{SR}(x, y) = \phi(y). \tag{2.13}$$

Since (M, d_{SR}) is complete, there exists a minimizing geodesic $\gamma : [0, 1] \to M$ between x and y. As before, we can parametrize the distribution Δ by a orthonormal family $\mathscr{F}$ of smooth vector fields $X^1, \ldots, X^m$ in a neighborhood $\mathscr{V}$ of $\gamma([0, 1])$, and we denote by u^γ the control corresponding to γ. By construction, it minimizes the quantity

$$C(u) = \int_0^1 \sum_{i=1}^m u_i(t)^2 dt,$$

among all the controls $u \in L^2([0, 1]; \mathbb{R}^m)$ which are admissible with respect to x, $\mathscr{F}$ and $\mathscr{V}$ and which satisfy the constraint $E_{\mathscr{F}}^{x,1}(u) = y$. Let $u \in L^2([0, 1]; \mathbb{R}^m)$ be a control admissible with respect to x, $\mathscr{F}$ and $\mathscr{V}$ such that $E_{\mathscr{F}}^{x,1}(u) \in \mathscr{U}$. By (2.13) one has

$$C(u) \geq e_{SR}\left(x, E_{\mathscr{F}}^{x,1}(u)\right) \geq \phi\left(E_{\mathscr{F}}^{x,1}(u)\right).$$

Moreover

$$C(u^\gamma) = e_{SR}(x, y) = \phi(y) = \phi\left(E_{\mathscr{F}}^{x,1}(u^\gamma)\right).$$

Hence u^γ minimizes the functional $D : L^2([0, 1]; \mathbb{R}^m) \to \mathbb{R}$ defined as

$$D(u) := C(u) - \phi\left(E_{\mathscr{F}}^{x,1}(u)\right),$$

over the set of controls $u \in L^2([0, 1]; \mathbb{R}^m)$ such that $E_{\mathscr{F}}^{x,1}(u) \in \mathscr{U}$. This means that u^γ is a critical point of D. Setting $\lambda = D_y\phi$, we obtain

$$\lambda \cdot D_{u^\gamma} E_{\mathscr{F}}^{x,1} - D_{u^\gamma} C = 0.$$

By Proposition 2.8, the path γ admits a normal extremal lift $\psi : [0, 1] \to T^*M$ satisfying $\psi(1) = (y, \frac{1}{2}D_y\phi)$. By the Cauchy-Lipschitz Theorem, such a normal extremal is unique. $\qquad \square$

Denote by $\mathscr{P}_x$ the set of points in M such that there is a unique normal minimizing geodesic γ_y from x to y. The previous lemma yields easily the following result.

Lemma 2.16 *The set $\mathscr{P}_x$ is dense in M.*

Proof (Proof of Lemma 2.16) Let $y \in M$ and $r > 0$ be fixed. Let $\varphi : M \to \mathbb{R}$ be a smooth function such that

$$\varphi(y) = 0 \quad \text{and} \quad \varphi(z) \geq 2r \quad \forall z \in \partial B_{SR}(y, r).$$

The continuous function

$$z \in \bar{B}_{SR}(x, r) \longmapsto d_{SR}(x, z) + \varphi(z)$$

is equal to $d_{SR}(x, y)$ at $z = y$ and by the triangle inequality it is larger than $d_{SR}(x, y)+r$ for $z \in \partial B_{SR}(y, r)$. Then there is $\bar{z} \in B_{SR}(y, r)$ such that

$$d_{SR}(x, z) \geq d_{SR}(x, \bar{z}) + \varphi(\bar{z}) - \varphi(z) \quad \forall z \in B_{SR}(y, r).$$

We conclude easily by Lemma 2.15.   □

For every $y \in \mathscr{P}_x$, denote by rank(y) the rank of the minimizing horizontal path γ_y (see Sect. 1.3).

Lemma 2.17 *The set of $y \in \mathscr{P}_x$ with rank(y) $= n$ is dense in M.*

Proof (Proof of Lemma 2.17) We argue by contradiction. Assume that there is an open set $\mathscr{O} \subset M$ such that any point $y \in \mathscr{P}_x \cap \mathscr{O}$ has rank $< n$. Set

$$\hat{r} := \max\left\{\text{rank}(y) \mid y \in \mathscr{P}_x \cap \mathscr{O}\right\}.$$

Fix $\hat{y} \in \mathscr{P}_x \cap \mathscr{O}$ such that rank($\hat{y}$) $= \hat{r}$ and set $\hat{\gamma} := \gamma_{\hat{y}}$. For every $y \in \mathscr{P}_x \cap \mathscr{O}$ denote by Π_y the affine subspace of $T_x^* M$ such that $\gamma_p = \gamma_y$, that is the space of $p \in T_x^* M$ such that

$$\gamma_p(t) = \exp_x(tp) = \pi\left(\psi_{x,p}(t)\right) = \gamma_y(t) \quad \forall t \in [0, 1].$$

Remembering Remark 2.4, we observe that the dimension of Π_y is exactly equal to $n - \text{rank}(y)$. As a matter of fact, given $y \in \mathscr{P}_x \cap \mathscr{O}$ and an orthonormal family $\mathscr{F} = \{X^1, \ldots, X^m\}$ in a neighborhood $\mathscr{V}$ along γ_y, remembering the arguments given in Proposition 2.8 we check that $p \in T_x^* M$ belongs to Π_y if and only if $\psi_{x,p}(1) = (y, p_p(1))$ satisfies

$$2p_p(1) \cdot D_{u^{\gamma_y}} E_{\mathscr{F}}^{x,1}(v) = D_{u^{\gamma_y}} C(v) \quad \forall v \in L^2([0, 1]; \mathbb{R}^m). \tag{2.14}$$

Let $\{y_k\}_k$ be a sequence in $\mathscr{P}_x \cap \mathscr{O}$ converging to $\hat{y}$, $\mathscr{F} = \{X^1, \ldots, X^m\}$ be an orthonormal family in a neighborhood $\mathscr{V}$ along $\hat{\gamma}$, and $\hat{u}$ the control associated with $\hat{\gamma}$ through $\mathscr{F}$. The End-Point mapping $E_{\mathscr{F}}^{x,1}$ is valued in $\mathbb{R}^n$; denote by $E_1, \ldots, E_n$ its n coordinates. The vector space (we identify $L^2([0, 1]; \mathbb{R}^m)$ with its dual)

$$\text{Span}\left\{D_{\hat{u}} E_1, \ldots, D_{\hat{u}} E_n\right\}$$

has dimension rank($\hat{y}$) $= \hat{r}$. Let $i_1, \ldots, i_{\hat{r}} \in \{1, \ldots, n\}$ be such that

$$\text{Span}\left\{D_{\hat{u}}E_{i_1}, \ldots, D_{\hat{u}}E_{i_{\hat{r}}}\right\} = \text{Span}\left\{D_{\hat{u}}E_1, \ldots, D_{\hat{u}}E_n\right\}. \tag{2.15}$$

Proceeding as in the proof of Proposition 2.3 and using completeness of (Δ, g), we show that taking a subsequence if necessary, $\{\gamma_k := \gamma_{y_k}\}_k$ converges uniformly to some minimizing geodesic joining x to y. By uniqueness, we infer that $\lim_{k \to +\infty} \gamma_k = \hat{\gamma}$. Furthermore, the proof also shows that the controls $u_k := u_{\gamma_k}$ which are associated to the γ_k's through the orthonormal family $\mathscr{F}$ converges strongly to $\hat{u}$ in $L^2([0, 1]; \mathbb{R}^m)$ (see Remark 2.1). Then by regularity of $E_{\mathscr{F}}^{x,1}$ and the fact that $\text{rank}(y_k) \leq \hat{r}$, we deduce that $\text{rank}(y_k) = \text{rank}(y)$ for k large enough and that

$$\text{Span}\left\{D_{u_k}E_{i_1}, \ldots, D_{u_k}E_{i_{\hat{r}}}\right\} = \text{Span}\left\{D_{u_k}E_1, \ldots, D_{u_k}E_n\right\}.$$

By (2.14) and (2.15), there is $\hat{\lambda} = (\hat{\lambda}_1, \ldots, \hat{\lambda}_{\hat{r}}) \in T_{\hat{y}}^* M \simeq (\mathbb{R}^{\hat{r}})^*$ such that (remember that we identify $L^2([0, 1]; \mathbb{R}^m)$ with its dual)

$$\sum_{j=1}^{\hat{r}} \hat{\lambda}_j D_{u^{\gamma_y}} E_{i_j} = u_{\gamma_y},$$

and more generally for every k there is $\lambda^k = (\lambda_1^k, \ldots, \lambda_{\hat{r}}^k) \in T_{y_k}^* M \simeq (\mathbb{R}^{\hat{r}})^*$ such that

$$\sum_{j=1}^{\hat{r}} \lambda_j^k D_{u_k} E_{i_j} = u_{\gamma_k},$$

Since $\{u_k\}_k$ converges to $\hat{u}$ in $L^2([0, 1]; \mathbb{R}^m)$, $\{D_{u_k}E_{\mathscr{F}}^{x,1}\}_k$ converges to $D_{\hat{u}}E_{\mathscr{F}}^{x,1}$ and $D_{\hat{u}}E_{i_1}, \ldots, D_{\hat{u}}E_{i_{\hat{r}}}$ are linearly independant, we infer that $\{\lambda^k\}_k$ tends to $\hat{\lambda}$ as k tends to $+\infty$. Define $\{p_k\}_k$ and $\hat{p}$ in $T_x^* M$ by

$$\psi_{x,p_k}(1) = (y_k, \lambda^k/2) \; \forall k \quad \text{and} \quad \psi_{x,\hat{p}}(1) = (\hat{y}, \hat{\lambda}/2).$$

By regularity of the Hamiltonian flow, $\{p_k\}_k$ tends to $\hat{p}$ and if a bounded sequence $\{p_k + q_k\}_k$ is contained in Π_k then it converges (up to a subsequence) to some point in $\Pi_{\hat{y}}$. This shows that Π_k tends to $\Pi_{\hat{y}}$. All in all we proved that the mapping $y \in \mathscr{P}_x \mapsto \Pi_y$ is continuous at $\hat{y}$. Let $\mathscr{S}$ be a smooth compact submanifold of dimension $\hat{r}$ in $T_x^* M$ which is transverse to $\Pi_{\hat{y}}$ at $\hat{p}$, that is such that

$$\Pi_{\hat{y}} \cap \mathscr{S} = \{\hat{p}\} \quad \text{and} \quad T_{\hat{p}}\mathscr{S} \cap T_{\hat{p}}\Pi_{\hat{y}} = \{0\}.$$

By regularity of $y \mapsto \Pi_y$, there is an open neighrborhood $\mathscr{O}' \subset \mathscr{O}$ of $\hat{y}$ such that $\mathscr{S}$ is transverse to any Π_y with $y \in \mathscr{P}_x \cap \mathscr{O}'$. We infer that

$$\{y\} = \exp_x\left(\Pi_y\right) = \exp_x\left(\Pi_y \cap \mathscr{S}\right) \subset \exp_x\left(\mathscr{S}\right) \qquad \forall y \in \mathscr{P}_x \cap \mathcal{O}'.$$

But since $\mathscr{S}$ has dimension strictly less than n, the set $\exp_x(\mathscr{S})$ is a compact set of measure zero in M. Then $\mathscr{P}_x \cap \mathcal{O}'$ cannot be dense in $\mathcal{O}'$. Which gives a contradiction. $\square$

Returning to the proof of Theorem 2.14, we fix $\bar{y} \in \mathscr{P}_x$ with rank$(\bar{y}) = n$. Given an open set $\Omega \subset M$, we call a function $f : \Omega \to \mathbb{R}$ *Lipschitz in charts* if it is Lipschitz in a set of local coordinates in a neighborhood of any point of Ω. This is equivalent to saying that f is locally Lipschitz with respect to a Riemannian distance on M.

Lemma 2.18 *There is an open set $\mathcal{O}_{\bar{y}}$ of $\bar{y}$ in M such that the function*

$$y \in \mathcal{O}_{\bar{y}} \longmapsto d_{SR}(x, y)$$

is Lipschitz in charts.

Proof (Proof of Lemma 2.18) As before we fix an orthonormal family of vector fields $\mathscr{F}$ in an open neighborhood $\mathcal{V}$ along $\bar{\gamma} := \gamma_{\bar{y}}$ which is associated with $\bar{u} \in L^2([0, 1]; \mathbb{R}^m)$ through $\mathscr{F}$. By a uniqueness-compactness argument, if $\{y_k\}_k$ converges to $\bar{y}$ and $\{\gamma_k\}_k$ is a sequence of minimizing geodesics between x and y_k then it converges (up to a subsequence) to $\bar{\gamma}$ and is associated with a sequence of controls $\{u_k\}_k$ which converges to $\bar{u}$ in $L^2([0, 1]; \mathbb{R}^m)$ (see Proposition 2.3 and Remark 2.1). Then there is a neighborhood $\mathcal{O}$ of $\bar{y}$ such that for every $y \in \mathcal{O}$ every minimizing geodesic between x and y is contained in $\mathcal{V}$ with rank n. Let $v^1, \dots v^n$ in $L^2([0, 1], \mathbb{R}^m)$ be such that the linear operator

$$\mathbb{R}^n \longrightarrow T_{\bar{y}}M$$
$$\alpha \longmapsto \sum_{i=1}^m \alpha_i D_{\bar{u}} E_{\mathscr{F}}^{x,1}\left(v^i\right)$$

is invertible. By continuity of $u \mapsto D_u E_{\mathscr{F}}^{x,1}$, taking $\mathcal{O}$ smaller if necessary, we may assume that for every $y \in \mathcal{O}$ and for every minimizing geodesic γ_y from x to y associated with a control u^y, the linear operator

$$\mathbb{R}^n \longrightarrow T_y M$$
$$\alpha \longmapsto \sum_{i=1}^m \alpha_i D_{u^y} E_{\mathscr{F}}^{x,1}\left(v^i\right)$$

is invertible. For every $y \in \mathcal{O}$, define $\mathscr{F}^y : \mathbb{R}^n \to M$ by

$$\mathscr{F}^y(\alpha) := E_{\mathscr{F}}^{x,1}\left(u^y + \sum_{i=1}^m \alpha_i v^i\right) \qquad \forall \alpha \in \mathbb{R}^n.$$

This mapping is well-defined and smooth in a neighborhood of the origin, satisfies

$$\mathscr{F}^y(0) = y,$$

and its differential at 0 is invertible. Hence by the Inverse Function Theorem, there are an open neighborhood $\mathscr{B}^y$ of y in M and a function $\mathscr{G}^y : \mathscr{B}^y \to \mathbb{R}^n$ with $\mathscr{G}^y(y) = 0$ such that

$$\mathscr{F}^y \circ \mathscr{G}^y(z) = z \qquad \forall z \in \mathscr{B}^y.$$

From the definition of the sub-Riemannian distance between two points, we infer that for any $z \in \mathscr{B}^y$ we have

$$d_{SR}(x, z) = \sqrt{e_{SR}(x, z)} \leq \left\| u^y + \sum_{i=1}^m \left(\mathscr{G}^y(z) \right)_i v^i \right\|_{L^2} =: \phi^y(z).$$

We conclude that, for every $y \in \mathcal{O}$, there are a open set $\mathscr{B}^y$ containing y and a C^1 function $\phi^y : \mathscr{B}^y \to \mathbb{R}^n$ such that

$$d_{SR}(x, y) = \phi^y(y) \quad \text{and} \quad d_{SR}(x, z) \leq \phi^y(z) \quad \forall z \in \mathscr{B}^y.$$

The C^1 norms of the ϕ^y's are uniformly bounded. This proves the lemma. $\qquad \square$

To conclude the proof of Theorem 2.14, we note that by the Rademacher Theorem, the function $y \in \mathcal{O}_{\bar{y}} \mapsto d_{SR}(x, y)$ is differentiable almost everywhere in $\mathcal{O}_{\bar{y}}$. By Lemma 2.15, for every $y \in \mathcal{O}_{\bar{y}}$ where the function is differentiable, there is $p_y \in T_x^* M$ such that

$$y = \exp_x(p_y), \quad d_{SR}(x, y) = \sqrt{2H(x, p_y)}, \quad \psi_{x, p_y}(1) = \left(y, \frac{1}{2} D_y d_{SR}^2(x, \cdot) \right).$$

Since $d_{SR}(x, \cdot)$ is Lipschitz in $\mathcal{O}_{\bar{y}}$, there is some constant $K > 0$ such that all the p_y's remain in a compact subset of $T_x^* M$. Now every $y \in \mathcal{O}_{\bar{y}}$ can be approximated by a sequence $\{y_k\}_k$ of points in $\mathcal{O}_{\bar{y}}$ where $d_{SR}(x, \cdot)$ is differentiable. By compactness, up to taking a subsequence, the normal extremals starting at (x, p_{y_k}) will converge to a normal extremal starting whose projection is a minimizing geodesic from x to y. $\qquad \square$

Remark 2.6 We already know that the sub-Riemannian distance is continuous on $M \times M$ (see Proposition 1.13). The proof of Theorem 2.14 shows that if (Δ, g) is complete and $x \in M$ be fixed, then the function $y \in M \to d_{SR}(x, y) \in \mathbb{R}$ is locally Lipschitz (in charts) on an open and dense subset of M.

2.4 The Goh Condition

Theorem 2.9 provides firt-order conditions for a given horizontal path to be a minimizing geodesic. The aim of the present section is to present a second-order necessary condition for a given singular path to be minimizing. For sake of simplicity, we fix an orthonormal family $\mathscr{F} = \{X^1, \ldots, X^m\}$ of smooth vector fields in some open chart $\mathscr{V}$ which contains a minimizing geodesic $\bar{\gamma} : [0, 1] \to M$ from x to y (with $x \neq y$). As before, we denote by $\bar{u} = u^{\bar{\gamma}}$ the control which is associated with $\bar{\gamma}$ through $\mathscr{F}$. Recall that $C \in L^2([0, 1]; \mathbb{R}^m)$ is defined by

$$C(u) := \|u\|_{L^2}^2 \qquad \forall u \in L^2([0, 1]; \mathbb{R}^m).$$

Define $F : L^2([0, 1]; \mathbb{R}^m) \to \mathbb{R}^n \times \mathbb{R}$ by

$$F(u) := \left(E_{\mathscr{F}}^{x,1}(u), C(u) \right) \qquad \forall u \in L^2([0, 1]; \mathbb{R}^m).$$

The Lagrange Multiplier Theorem asserts that if $\bar{u}$ minimizes $C(u)$ under the constraint $E_{\mathscr{F}}^{x,1}(u) = y$, then there are $\lambda \in (\mathbb{R}^n)^*$ and $\lambda_0 \in \{0, 1\}$ with $(\lambda, \lambda_0) \neq (0, 0)$ such that

$$\lambda \cdot D_{\bar{u}} E_{\mathscr{F}}^{x,1} = \lambda_0 D_{\bar{u}} C.$$

In Sect. 2.2, we saw that whenever $\lambda_0 = 0$ we cannot deduce that $\bar{\gamma}$ satisfies the geodesic equation, that is that it is the projection of a normal extremal. In the case $\lambda_0 = 0$, the control $\bar{u} \in \mathscr{U}_{\mathscr{F}}^{x,1}$ is necessarily singular which means that it is a critical point of $E_{\mathscr{F}}^{x,1}$. Thus we have to study what happens at second order.

Let $\mathscr{U}$ be an open set in $L^2 = L^2([0, 1]; \mathbb{R}^m)$ and $F : \mathscr{U} \to \mathbb{R}^N$ be a function of class C^2 with respect to the L^2-norm. We recall that we call *critical point* of F any $u \in \mathscr{U}$ such that $D_u F : \mathscr{U} \to \mathbb{R}^N$ is not surjective. Given a critical point u, we call corank of u, the quantity

$$\mathrm{corank}_F(u) := N - \dim\left(\mathrm{Im}(D_u F)\right).$$

For every $u \in \mathscr{U}$ the second differential of F at u is the quadratic mapping on $D_u^2 F : L^2 \to \mathbb{R}^N$ satisfying

$$F(u + v) = F(u) + D_u F(v) + \frac{1}{2} D_u^2 F \cdot (v, v) + \|v\|_{L^2}^2 \, o(1).$$

If $Q : L^2 \to \mathbb{R}$ is a quadratic form, we define its *negative index* by

$$\mathrm{ind}_-(Q) := \max\left\{\dim(L) \mid Q_{|L \setminus \{0\}} < 0\right\}.$$

We are now ready to state the result whose proof is given in Appendix B.

Theorem 2.19 *Let $F : \mathscr{U} \to \mathbb{R}^N$ be a mapping of class C^2 in an open set $\mathscr{U} \subset L^2$ and $\bar{u} \in \mathscr{U}$ be a critical point of F of corank r. If*

$$ind_- \left(\lambda^* \left(D_{\bar{u}}^2 F \right)_{|Ker(D_{\bar{u}}F)} \right) \geq r \quad \forall \lambda \in \left(Im(D_{\bar{u}}F) \right)^\perp \setminus \{0\},$$

then the mapping F is locally open at $\bar{u}$, that is the image of any neighborhood of $\bar{u}$ is an neighborhood of $F(\bar{u})$.

From Proposition 1.11 and Remark 1.8, we know that for every non-zero form $\bar{p} \in (\mathbb{R}^n)^*$ with

$$\bar{p} \cdot D_{\bar{u}} E_{\mathscr{F}}^{x,1} = 0,$$

the absolutely continuous arc $\bar{p} : [0, 1] \to (\mathbb{R}^n)^* \setminus \{0\}$ defined by

$$\bar{p}(t) := \bar{p} \cdot \bar{S}(1)\bar{S}(t)^{-1} \quad \forall t \in [0, 1], \tag{2.16}$$

satisfies $\bar{p}(1) = \bar{p}$,

$$\dot{\bar{p}}(t) = - \sum_{i=1}^k \bar{u}_i(t) \bar{p}(t) \cdot D_{\gamma_{\bar{u}}(t)} X^i \quad \text{a.e. } t \in [0, T],$$

and

$$\bar{p}(t) \cdot X^i \left(\gamma_{\bar{u}}(t) \right) = 0 \quad \forall t \in [0, T], \forall i = 1, \ldots m,$$

where $\bar{S} : [0, T] \to M_n(\mathbb{R})$ is the solution to the Cauchy problem

$$\dot{\bar{S}}(t) = \bar{A}(t)\bar{S}(t) \quad \text{a.e. } t \in [0, T], \quad \bar{S}(0) = I_n, \tag{2.17}$$

and the matrices $\bar{A}(t) \in M_n(\mathbb{R})$, $\bar{B}(t) \in M_{n,k}(\mathbb{R})$ are defined by

$$\bar{A}(t) := \sum_{i=1}^m \bar{u}_i(t) J_{X^i} \left(\gamma_{\bar{u}}(t) \right) \quad \text{a.e. } t \in [0, T] \tag{2.18}$$

and

$$\bar{B}(t) := \left(X^1(\gamma_{\bar{u}}(t)), \cdots, X^m(\gamma_{\bar{u}}(t)) \right) \quad \forall t \in [0, T]. \tag{2.19}$$

The following result combined with Theorem 2.19 will yield a necessary condition for a minimizing horizontal path to be strictly abnormal. It holds in the general case of a control $\bar{u}$ which belongs to $\mathscr{U}_{\mathscr{F}}^{x,1} \cap L^\infty([0, 1]; \mathbb{R}^m)$. We do not need $\bar{u}$ to be minimizing.

Theorem 2.20 Let $\bar{u} \in \mathscr{U}_{\mathscr{F}}^{x,1} \cap L^\infty([0, 1]; \mathbb{R}^m)$ and $\bar{p} \in (\mathbb{R}^n)^* \setminus \{0\}$ be such that

$$\bar{p} \cdot D_{\bar{u}} E_{\mathscr{F}}^{x,1} = 0. \tag{2.20}$$

Assume that

$$ind_- \left(\bar{p} \cdot \left(D_{\bar{u}}^2 E_{\mathscr{F}}^{x,1} \right)_{| \, Ker(D_{\bar{u}} E_{\mathscr{F}}^{x,1})} \right) < +\infty. \tag{2.21}$$

Then the absolutely continuous arc $\bar{p} : [0, 1] \to (\mathbb{R}^n)^* \setminus \{0\}$ defined by (2.16) satisfies

$$\bar{p}(t) \cdot \left[X^i, X^j \right] \left(\gamma_{\bar{u}}(t) \right) = 0 \quad \forall t \in [0, 1], \; \forall i, j = 1, \dots, m. \tag{2.22}$$

Proof Let us first check that F is of class C^2 on $\mathscr{U}_{\mathscr{F}}^{x,1}$ (we refer the reader to Appendix B for basics in differential calculus in infinite dimension). Given $u \in U_{\mathscr{F}}^{x,1}$ and $v \in L^2([0, T]; \mathbb{R}^m)$ we need to study the quantity

$$\gamma_{u+\varepsilon v}(1) - \gamma_u(1) = E_{\mathscr{F}}^{x,1} \left(u + \varepsilon v \right) - E_{\mathscr{F}}^{x,1} \left(u \right),$$

at second order when ε is small. We have

$$\gamma_{u+\varepsilon v}(1) = \int_0^1 \sum_{i=1}^k \left(u_i(t) + \varepsilon v_i(t) \right) X^i \left(\gamma_{u+\varepsilon v}(t) \right) dt, \tag{2.23}$$

with $\gamma_{u+\varepsilon v}(0) = x$. For every $i = 1, \dots, m$ and every $t \in [0, 1]$, the Taylor expansion of each X^i at $\gamma_u(t)$ at second order gives

$$
\begin{aligned}
X^i \left(\gamma_{u+\varepsilon v}(t) \right) ={} & X^i \left(\gamma_u(t) \right) + D_{\gamma_u(t)} X^i \cdot \left(\gamma_{u+\varepsilon v}(t) - \gamma_u(t) \right) \\
& + \frac{1}{2} D_{\gamma_u(t)}^2 X^i \cdot \left(\gamma_{u+\varepsilon v}(t) - \gamma_u(t), \gamma_{u+\varepsilon v}(t) - \gamma_u(t) \right) \\
& + \left| \gamma_{u+\varepsilon v}(t) - \gamma_u(t) \right|^2 o(1).
\end{aligned}
$$

Setting $\delta_x(t) := \gamma_{u+\varepsilon v}(t) - \gamma_u(t)$ for any t, (2.23) yields formally (δ_x has size ε)

$$
\begin{aligned}
\delta_x(1) ={} & \int_0^1 \left[\sum_{i=1}^m u_i(t) D_{\gamma_u(t)} X^i \cdot \delta_x(t) + \varepsilon \sum_{i=1}^m v_i(t) X^i \left(\gamma_u(t) \right) \right] dt \\
& + \int_0^1 \left[\varepsilon \sum_{i=1}^m v_i(t) D_{\gamma_u(t)} X^i \cdot \delta_x(t) + \frac{1}{2} \sum_{i=1}^m u_i(t) D_{\gamma_u(t)}^2 X^i \cdot (\delta_x(t), \delta_x(t)) \right] dt \\
& + \|v\|_\infty^2 o(1).
\end{aligned}
$$

Writing $\delta_x(t)$ as $\delta_x(t) = \delta_x^1(t) + \delta_x^2(t) + o(\varepsilon^2)$ where δ_x^1 is linear in ε and δ_x^2 is quadratic in ε we infer that δ_x^1 and δ_x^2 must satisfy formally

$$\dot{\delta}_x^1(t) = \left[\sum_{i=1}^m u_i(t)D_{\gamma_u(t)}X^i\right] \cdot \delta_x^1(t) + \left[\varepsilon \sum_{i=1}^k v_i(t)X^i\big(\gamma_u(t)\big)\right] \qquad \text{a.e. } t \in [0,1]$$

and

$$\dot{\delta}_x^2(t) = \left[\sum_{i=1}^m u_i(t)D_{\gamma_u(t)}X^i\right] \cdot \delta_x^2(t) + \left[\varepsilon \sum_{i=1}^m v_i(t)D_{\gamma_u(t)}X^i\right] \cdot \delta_x^1(t)$$
$$+ \frac{1}{2}\sum_{i=1}^m u_i(t)D^2_{\gamma_u(t)}X^i \cdot \big(\delta_x^1(t), \delta_x^1(t)\big) \qquad \text{a.e. } t \in [0,1].$$

Then from the Taylor expansion

$$\gamma_{u+\varepsilon v}(1) = \gamma_u(1) + \delta_x^1(1) + \delta_x^2(1) + o\big(\varepsilon^2\big),$$

we obtain that (here we use the notations of the proof of Proposition 1.8)

$$D_u^2 E_{\mathscr{F}}^{x,1} \cdot (v,v) = 2\int_0^1 S(1)S(t)^{-1}\left[C(t) + D(t)\right] dt \qquad \forall v \in L^2\big([0,1]; \mathbb{R}^m\big)$$

where

$$C(t) = \sum_{i=1}^m v_i(t)D_{\gamma_u(t)}X^i \cdot \delta_x^1(t) \tag{2.24}$$

and

$$D(t) = \frac{1}{2}\sum_{i=1}^m u_i(t)D^2_{\gamma_u(t)}X^i \cdot \big(\delta_x^1(t), \delta_x^1(t)\big). \tag{2.25}$$

Using Gronwall's lemma, we leave the reader to check that $E_{\mathscr{F}}^{x,1}$ is C^2 on $\mathscr{U}_{\mathscr{F}}^{x,1}$.

Let us now fix $\bar{u} \in \mathscr{U}_{\mathscr{F}}^{x,1} \cap L^\infty\big([0,1]; \mathbb{R}^m\big)$ and $\bar{p} \in (\mathbb{R}^n)^* \setminus \{0\}$ such that (2.20) and (2.21) are satisfied and prove that (2.22) holds. Note that we have for every $v \in L^2\big([0,1]; \mathbb{R}^m\big)$,

$$D_{\bar{u}}^2 E_{\mathscr{F}}^{x,1} \cdot (v,v) = 2\int_0^1 \bar{S}(1)\bar{S}(t)^{-1}\left[\bar{C}(t) + \bar{D}(t)\right] dt, \tag{2.26}$$

where $\bar{C}, \bar{D}$ are obtained by replacing u by $\bar{u}$ in (2.24)–(2.25) and the definitions of $\bar{S}, \bar{A}, \bar{B}$ (see (2.17)–(2.19)).

Lemma 2.21 *There is $K > 0$ such that for any $\bar{t}, \delta > 0$ with $[\bar{t}, \bar{t} + \delta] \subset [0,1]$, there holds for every $v \in \mathrm{Ker}\big(D_{\bar{u}}E_{\mathscr{F}}^{x,1}\big)$ with $\mathrm{Supp}(v) \in [\bar{t}, \bar{t} + \delta]$,*

$$\left| D_{\bar{u}}^2 E_{\mathscr{F}}^{x,1} \cdot (v, v) - \bar{Q}_{\bar{t}, \delta}(v) \right| \le K \, \|v\|_{L^2}^2 \, \delta^2,$$

where $\bar{Q}_{\bar{t}, \delta} : L^2 \left([0, 1]; \mathbb{R}^m \right) \to \mathbb{R}^n$ is defined by

$$\bar{Q}_{\bar{t}, \delta}(v) := \int_{\bar{t}}^{\bar{t}+\delta} \bar{p}(\bar{t}) \cdot \sum_{i=1}^m v_i(t) D_{\bar{\gamma}(\bar{t})} X^i \left[\int_{\bar{t}}^t \sum_{j=1}^m v_j(s) X^j \left(\bar{\gamma}(\bar{t}) \right) ds \right] dt, \qquad (2.27)$$

for every $v \in L^2 \left([0, 1]; \mathbb{R}^m \right)$.

Proof (Proof of Lemma 2.21) Let $\bar{t}, \delta > 0$ with $[\bar{t}, \bar{t} + \delta] \subset [0, 1]$ and $v \in \text{Ker}(D_{\bar{u}} E_{\mathscr{F}}^{x,1})$ with $\text{Supp}(v) \in [\bar{t}, \bar{t} + \delta]$ be fixed. By Remark 1.5, we have

$$\bar{S}(1) \int_0^1 \bar{S}(t)^{-1} \bar{B}(t) v(t) \, dt = 0.$$

Then (2.26) yields

$$\bar{p} \cdot \left(D_{\bar{u}}^2 E_{\mathscr{F}}^{x,1} \right)_{|\text{Ker}(d_{\bar{u}} E_{\mathscr{F}}^{x,1})} (v) = 2 \int_0^1 \bar{p}(t) \cdot \left[\bar{C}(t) + \bar{D}(t) \right] dt.$$

Setting

$$\bar{\delta}_x^1(t) := \bar{S}(t) \int_0^t \bar{S}(s)^{-1} \bar{B}(s) v(s) \, ds \qquad \forall t \in [0, 1],$$

we have $\bar{\delta}_x^1(t) = 0$ for every $t \in [0, \bar{t}] \cup [\bar{t} + \delta, 1]$ and by Cauchy-Schwarz's inequality, we have for every $t \in [\bar{t}, \bar{t} + \delta]$,

$$\left| \bar{\delta}_x^1(t) \right| \le \sup_{s \in [0, 1]} \left\{ \left\| \bar{S}(t) \bar{S}(s)^{-1} \bar{B}(s) \right\| \right\} \sqrt{t - \bar{t}} \, \|v\|_{L^2} \le K_1 \sqrt{\delta} \, \|v\|_{L^2},$$

where K_1 is a constant depending only upon the sizes of $\bar{S}, \bar{S}^{-1}, \bar{B}$ in a neighborhood of the curve $\gamma_{\bar{u}} ([0, 1])$. Then we have

$$\bar{D}(t) = 0 \qquad \forall t \in t \in [0, \bar{t}] \cup [\bar{t} + \delta, 1],$$

and

$$\left| \bar{D}(t) \right| \le K_3 \, \delta \, \|v\|_{L^2}^2 \, \|\bar{u}\|_{L^\infty} \qquad \forall t \in [\bar{t}, \bar{t} + \delta],$$

which gives

$$\left| \int_0^1 \bar{p}(t) \cdot \bar{D}(t) \, dt \right| \le K_4 \, \|v\|_{L^2}^2 \, \delta^2,$$

where K_3, K_4 are some constants depending on K_1, on the size of the $D^2 X^j$'s, $\bar{p}$ and $\|u\|_{L^\infty}$. Note that since we can write ($\bar{\gamma} = \gamma_{\bar{u}}$)

$$\bar{\delta}_x^1(t) - \int_{\bar{t}}^t \sum_{j=1}^m v_j(s) X^j(\bar{\gamma}(\bar{t})) \, ds$$

$$= \bar{\delta}_x^1(t) - \int_{\bar{t}}^t \sum_{j=1}^m v_j(s) X^j(\bar{\gamma}(s)) \, ds + \int_{\bar{t}}^t \sum_{j=1}^m v_j(s) \left[X^j(\bar{\gamma}(s)) - X^j(\bar{\gamma}(\bar{t})) \right] ds$$

$$= \int_0^t \bar{S}(t) \bar{S}(s)^{-1} \bar{B}(s) v(s) - \bar{B}(s) v(s) \, ds + \int_{\bar{t}}^t \sum_{j=1}^m v_j(s) \left[X^j(\bar{\gamma}(s)) - X^j(\bar{\gamma}(\bar{t})) \right] ds$$

$$= \int_{\bar{t}}^t \left(\bar{S}(t) - \bar{S}(s) \right) \bar{S}(s)^{-1} \bar{B}(s) v(s) \, ds + \int_{\bar{t}}^t \sum_{j=1}^m v_j(s) \left[X^j(\bar{\gamma}(s)) - X^j(\bar{\gamma}(\bar{t})) \right] ds,$$

we have (since $\bar{u}$ belongs to $L^\infty([0, 1]; \mathbb{R}^m)$, $\bar{S}$ and $\bar{\gamma}$ are both Lipschitz)

$$\left| \bar{\delta}_x^1(t) - \int_{\bar{t}}^t \sum_{j=1}^m v_j(s) X^j(\bar{\gamma}(\bar{t})) \, ds \right| \leq K_2 \|v\|_{L^2} \delta^{\frac{3}{2}}, \quad \forall t \in t \in [\bar{t}, \bar{t} + \delta], \quad (2.28)$$

where K_1 is a constant depending only upon the sizes of $\bar{S}$, $\bar{S}^{-1}$, $\bar{B}$ and the Lipschitz constants of the X^j's in a neighborhood of the curve $\gamma_{\bar{u}}([0, 1])$. By (2.27), we have

$$\int_0^1 \bar{p}(t) \cdot \bar{C}(t) \, dt - \bar{Q}_{\bar{t}, \delta}(v) = \int_{\bar{t}}^{\bar{t}+\delta} \bar{p}(t) \cdot \bar{C}(t) \, dt - \bar{Q}_{\bar{t}, \delta}(v)$$

$$= \int_{\bar{t}}^{\bar{t}+\delta} \bar{p}(t) \cdot \left(\sum_{i=1}^m v_i(t) D_{\bar{\gamma}(t)} X^i \cdot \bar{\delta}_x^1(t) - \sum_{i=1}^m v_i(t) D_{\bar{\gamma}(\bar{t})} X^i \cdot \left[\int_{\bar{t}}^t \sum_{j=1}^m v_j(s) X^j(\bar{\gamma}(\bar{t})) \, ds \right] \right) dt$$

$$= \int_{\bar{t}}^{\bar{t}+\delta} \bar{p}(t) \cdot \left(\sum_{i=1}^m v_i(t) D_{\bar{\gamma}(t)} X^i \right) \cdot \left[\bar{\delta}_x^1(t) - \int_{\bar{t}}^t \sum_{j=1}^m v_j(s) X^j(\bar{\gamma}(\bar{t})) \, ds \right] dt.$$

By (2.28), we infer that

$$\left| \int_{\bar{t}}^{\bar{t}+\delta} \bar{p}(t) \cdot \bar{C}(t) \, dt - \bar{Q}_{\bar{t}, \delta}(v) \right| \leq K_5 \|v\|_{L^2}^2 \delta^2,$$

for some constant K_5 depending on the datas. All in all, we get

$$\left| \int_0^1 \bar{p}(t) \cdot \left[\bar{C}(t) + \bar{D}(t) \right] dt - \bar{Q}_{\bar{t}, \delta}(v) \right| \leq K_6 \|v\|_{L^2}^2 \delta^2,$$

for some constant K_6 depending on the datas. We conclude easily. $\qquad\square$

Returning to the proof of Theorem 2.20, we argue by contradiction and assume that (2.22) does not hold. Hence we assume that there are $\bar{t} \in (0, 1)$ and $\bar{i} \neq \bar{j} \in \{1, \cdots, m\}$ such that

$$N_{\bar{i}, \bar{j}}(\bar{t}) := \bar{p}(\bar{t}) \cdot \left[X^{\bar{i}}, X^{\bar{j}}\right] (\bar{\gamma}(\bar{t})) > 0$$
$$= \bar{p}(\bar{t}) \cdot \left(D_{\bar{\gamma}(\bar{t})} X^{\bar{j}} \cdot X^{\bar{i}}(\bar{\gamma}(\bar{t})) - D_{\bar{\gamma}(\bar{t})} X^{\bar{i}} \cdot X^{\bar{j}}(\bar{\gamma}(\bar{t}))\right).$$

Let $\delta > 0$ such that $[\bar{t}, \bar{t} + \delta] \subset [0, 1]$ and $\bar{Q}_{\bar{t}, \delta} : L^2([0, 1] \to \mathbb{R}^n$ be the mapping defined by (2.27). We observe that there holds for every $v \in L^2([0, 1]; \mathbb{R}^m)$,

$$\bar{Q}_{\bar{t}, \delta}(v) = \int_{\bar{t}}^{\bar{t}+\delta} \int_{\bar{t}}^{t} \left[\sum_{i,j=1}^{m} v_i(t) v_j(s) \left(\bar{p}(\bar{t}) \cdot D_{\bar{\gamma}(\bar{t})} X^i \cdot X^j(\bar{\gamma}(\bar{t}))\right)\right] ds\, dt$$
$$= \int_{\bar{t}}^{\bar{t}+\delta} \int_{\bar{t}}^{t} \langle v(s), \bar{M}v(t)\rangle\, ds\, dt = \int_{\bar{t}}^{\bar{t}+\delta} \langle w(t), \bar{M}v(t)\rangle\, dt,$$

where $\bar{M}$ is the $m \times m$ matrix defined by

$$\bar{M}_{i,j} = \bar{p}(\bar{t}) \cdot D_{\bar{\gamma}(\bar{t})} X^i \cdot X^j(\bar{\gamma}(\bar{t})),$$

and

$$w(t) := \int_{\bar{t}}^{t} v(s)\, ds \qquad \forall t \in [0, 1].$$

Thanks to Lemma 2.21, in order to get a contradiction, we need to show that for every integer $N > 0$, there are $\delta > 0$ and a subspace $L_\delta \subset L^2([0, 1]; \mathbb{R}^m)$ of dimension larger than N such that the restriction of $\bar{Q}_{\bar{t}, \delta}$ to $L \setminus \{0\}$ satisfies the following property:

$$\bar{Q}_{\bar{t}, \delta}(v) < -K \|v\|_{L^2}^2 \delta^2 \qquad \forall v \in L \setminus \{0\}.$$

As a matter of fact, given $N \in \mathbb{N}$ strictly larger than n, if L is a vector subspace of dimension N, then the linear operator

$$\left(D_{\bar{u}} E_{\mathscr{F}}^{x,1}\right)_{|L} : L \longrightarrow \mathbb{R}^n$$

has a kernel of dimension at least $N - n$, which means that

$$\mathrm{Ker}\left(D_{\bar{u}} E_{\mathscr{F}}^{x,1}\right) \cap L$$

has dimension at least $N - n$.

Let N an integer strictly larger than n be fixed and $\delta > 0$ with $[\bar{t}, \bar{t} + \delta] \subset [0, 1]$ to be chosen later. Denote by $L = L_{\bar{t}, \delta, N}$ the vector space in $L^2([0, 1]; \mathbb{R}^m)$ of all the controls v such that there is a sequence $\{a_1, \ldots, a_N\}$ such that

$$\begin{cases} v_{\bar{i}}(t) = \sum_{k=1}^{N} a_k \cos\left(k \frac{(t-\bar{t})2\pi}{\delta}\right) \\ v_{\bar{j}}(t) = \sum_{k=1}^{N} a_k \sin\left(k \frac{(t-\bar{t})2\pi}{\delta}\right) \end{cases} \quad \forall t \in [\bar{t}, \bar{t} + \delta],$$

$$v_{\bar{i}}(t) = v_{\bar{j}}(t) = 0 \quad \forall t \notin [\bar{t}, \bar{t} + \delta],$$

and

$$v_i(t) = 0, \quad \forall i \neq \bar{i}, \bar{j} \quad \forall t \in [0, 1].$$

Let $v \in L \setminus \{0\}$, taking as before $w(t) := \int_{\bar{t}}^{t} v(s)ds$, we have

$$\begin{cases} w_{\bar{i}}(t) = \frac{\delta}{2\pi} \sum_{k=1}^{N} \frac{a_k}{k} \sin\left(k \frac{(t-\bar{t})2\pi}{\delta}\right) \\ w_{\bar{j}}(t) = \frac{\delta}{2\pi} \sum_{k=1}^{N} \frac{a_k}{k}\left(1 - \cos\left(k \frac{(t-\bar{t})2\pi}{\delta}\right)\right), \end{cases} \quad \forall t \in [\bar{t}, \bar{t} + \delta],$$

$$w_{\bar{i}}(t) = w_{\bar{j}}(t) = 0 \quad \forall t \notin [\bar{t}, \bar{t} + \delta],$$

and

$$w_i(t) = 0, \quad \forall i \neq \bar{i}, \bar{j} \quad \forall t \in [0, 1].$$

Then we have

$$\int_0^1 w_{\bar{i}}(t)v_{\bar{j}}(t)\, dt = \sum_{k=1}^{+\infty} \frac{\delta^2 a_k^2}{4\pi k}$$

and

$$\int_0^1 w_{\bar{j}}(t)v_{\bar{i}}(t)\, dt = -\sum_{k=1}^{+\infty} \frac{\delta^2 a_k^2}{4\pi k}.$$

We have for every $t \in [0, 1]$

$$\langle w(t), \bar{M}v(t)\rangle = w_{\bar{i}}(t)\bar{M}_{\bar{i}\bar{i}}v_{\bar{i}}(t) + w_{\bar{i}}(t)\bar{M}_{\bar{i}\bar{j}}v_{\bar{j}}(t) + w_{\bar{j}}(t)\bar{M}_{\bar{j}\bar{i}}v_{\bar{i}}(t) + w_{\bar{j}}(t)\bar{M}_{\bar{j}\bar{j}}v_{\bar{j}}(t).$$

But

$$\int_0^1 w_{\bar{i}}(t)\bar{M}_{\bar{i}\bar{i}}v_{\bar{i}}(t)\,dt = \bar{M}_{\bar{i}\bar{i}}\int_0^1 w_{\bar{i}}(t)\dot{w}_{\bar{i}}(t)\,dt = 0 = \int_0^1 w_{\bar{j}}(t)\bar{M}_{\bar{j}\bar{j}}v_{\bar{j}}(t)\,dt.$$

In conclusion, we have

$$\bar{Q}_{\bar{i},\delta}(v) = \int_0^1 \langle w(t), \bar{M}v(t)\rangle\,dt = -N_{\bar{i},\bar{j}}(\bar{t})\sum_{k=1}^N \frac{\delta^2 a_k^2}{4\pi k} = -\frac{\delta^2 N_{\bar{i},\bar{j}}(\bar{t})}{4\pi}\sum_{k=1}^N \frac{a_k^2}{k}.$$

Since $N_{\bar{i},\bar{j}}(\bar{t}) > 0$, $\bar{Q}_{\bar{i},\delta}(v)$ is negative. Moreover, we observe that

$$\|v\|_{L^2}^2 = \delta\sum_{k=1}^N a_k^2,$$

which yields

$$\frac{|\bar{Q}_{\bar{i},\delta}(v)|}{\|v\|_{L^2}^2\delta^2} = \frac{N_{\bar{i},\bar{j}}(\bar{t})}{4\pi}\frac{\sum_{k=1}^N \frac{a_k^2}{k}}{\delta\sum_{k=1}^N a_k^2} \geq \frac{1}{\delta}\left(\frac{N_{\bar{i},\bar{j}}(\bar{t})}{4\pi}\right).$$

We conclude easily by taking $\delta > 0$ small enough.   □

A minimizing geodesic is called *strictly abnormal* if it is singular and admits no normal extremal lift. A control is called *strictly abnormal* if its associated horizontal path is strictly abnormal.

Theorem 2.22 *Let $\bar{\gamma} : [0, 1] \to M$ be a minimizing geodesic from x to y (with $x \neq y$) which is strictly abnormal. Then there is an abnormal lift $\bar{\psi} = (\bar{\gamma}, \bar{p}) : [0, 1] \to T^*M$ of $\bar{\gamma}$ such that*

$$\bar{p}(t)\cdot[X^i, X^j](\bar{\gamma}(t)) = 0 \quad \forall t \in [0, 1], \forall i,j = 1,\dots,m.$$

The latter property is called the Goh condition and $\bar{\gamma}$ is a Goh path.

Proof According to the previous notations, we define the mapping $F : \mathscr{U}_{\mathscr{F}}^{x,1} \to \mathbb{R}^n \times \mathbb{R}$ by

$$F(u) := \left(E_{\mathscr{F}}^{x,1}(u), \|u\|_{L^2}^2\right) \quad \forall u \in \mathscr{U}_{\mathscr{F}}^{x,1}.$$

This function, which is of class C^2, cannot be open at $\bar{u}$. As a matter of fact, if the image of a neighborhood of $\bar{u}$ contains a neighborhood of $F(\bar{u})$ then it contains a control $u \in \mathscr{U}_{\mathscr{F}}^{x,1}$ with

$$E_{\mathscr{F}}^{x,1}(u) = y \quad \text{and} \quad \|u\|_{L^2}^2 \leq \|\bar{u}\|_{L^2}^2,$$

which contradicts the minimality of $\bar{u}$ from x to y. Therefore by Theorem 2.19 we infer that there is $\lambda \in (\text{Im}(D_{\bar{u}}F))^{\perp}\setminus\{0\}$ such that

$$\text{ind}_- \left(\lambda^* \left(D_{\bar{u}}^2 F \right)_{|\text{Ker}(d_{\bar{u}} F)} \right) < r := n - \text{rank}(\bar{u}).$$

Since the control $\bar{u}$ is strictly abnormal, the last coordinates of λ is zero. Denote by $\bar{p}$ the dual of the first n coordinates of λ. Then we have

$$\bar{p} \cdot D_{\bar{u}} E_{\mathscr{F}}^{x,1}(v) = 0 \qquad \forall v \in L^2([0,1]; \mathbb{R}^m).$$

Since $\bar{u}$ is minimizing, $|u(t)|$ is constant and $\bar{u}$ belongs to $L^\infty([0,1]; \mathbb{R}^m)$. Theorem 2.20 concludes the proof. $\square$

Example 2.1 A distribution Δ is called *medium-fat* if, for every $x \in M$ and every section X of Δ with $X(x) \neq 0$, there holds

$$T_x M = \Delta(x) + [\Delta, \Delta](x) + \big[X, [\Delta, \Delta] \big](x), \tag{2.29}$$

where

$$[\Delta, \Delta](x) := \Big\{ [X,Y](x) \,|\, X, Y \text{ sections of } \Delta \Big\}$$

$$\text{and} \quad \big[X, [\Delta, \Delta] \big](x) := \Big\{ \big[X, [Y,Z] \big](x) \,|\, Y, Z \text{ sections of } \Delta \Big\}.$$

Any two-generating distribution is medium-fat. An example of medium-fat distribution which is not two-generating is given by the rank-three distribution in $\mathbb{R}^4$ with coordinates $x = (x_1, x_2, x_3, x_4)$ generated by the vector fields defined by

$$X^1 = \partial_{x_1}, \quad X^2 = \partial_{x_2}, \quad X^3 = \partial_{x_3} + (x_1 + x_2 + x_3)^2 \partial_{x_4}.$$

Medium-fat distribution do not admit non-trivial Goh paths. As a matter of fact, if $\gamma : [0,T] \to M$ is an horizontal path which admits an abnormal lift $\psi = (\gamma, p) : [0,T] \to T^*M$ satisfying the Goh condition, then we have

$$p(t) \cdot \big[X^i, X^j \big](\gamma(t)) = 0 \qquad \forall i, j = 1, \ldots, m, \tag{2.30}$$

for every t in a small interval $I \subset [0,T]$ such that $\gamma(t)$ is in a local chart of M and Δ is parametrized by a family $\{X^1, \ldots, X^m\}$ of smooth vector fields. Then if we denote by u the control which is associated to γ through $\mathscr{F}$, derivating the previous equality yields for any $i, j = 1, \ldots, m$,

$$p(t) \cdot \left[\sum_{k=1}^m u_k(t) X^k, \big[X^i, X^j \big] \right] (\gamma(t)) = 0 \qquad \forall t \in I. \tag{2.31}$$

Since $\psi = (\gamma, p)$ is an abnormal lift, we also have $p \cdot X^i = 0$ along γ, then by (2.29), (2.30) and (2.31) we get a contradiction.

2.5 Examples of SR Geodesics

Geodesics in the Heisenberg group. The *Heisenberg group* $\mathbb{H}^1$ is the sub-Riemannian structure (Δ, g) in $\mathbb{R}^3$ where Δ is the totally nonholonomic rank 2 distribution (see Example 1.2) spanned by the vector fields

$$X = \partial_x - \frac{y}{2}\partial_z \quad \text{and} \quad Y = \partial_y + \frac{x}{2}\partial_z,$$

and g is the metric making the family $\{X, Y\}$ orthonormal, that is defined by

$$g = dx^2 + dy^2.$$

The above structure can be shown to be left-invariant under the group law

$$(x, y, z) \star (x', y', z') = \left(x + x', y + y', z + z' + \frac{1}{2}(xy' - x'y)\right).$$

Thanks to Proposition 1.7, any horizontal path on $[0, T]$ has the form $\gamma_u = (x, y, z) : [0, T] \to \mathbb{R}^3$ where

$$\begin{cases} \dot{x}(t) = u_1(t) \\ \dot{y}(t) = u_2(t) \\ \dot{z}(t) = \frac{1}{2}(u_2(t)x(t) - u_1(t)y(t)), \end{cases}$$

for some $u \in L^2([0, T]; \mathbb{R}^2)$. This means that

$$z(T) - z(0) = \int_0^T \frac{1}{2}(x(t)\dot{y}(t) - y(t)\dot{x}(t))\, dt = \int_c \frac{1}{2}(xdy - ydx),$$

where $\alpha(t) = (x(t), y(t))$ is the projection of the curve γ to the plane. According to the Stockes Theorem, we have

$$\int_\alpha \frac{1}{2}(xdy - ydx) = \int_{\mathscr{D}} dx \wedge dy + \int_c \frac{1}{2}(xdy - ydx),$$

where $\mathscr{D}$ denotes the domain which is enclosed by the curve α and the segment

$$c := [\varrho_1, \varrho_2] := [(x(0), y(0)), (x(T), y(T))]$$

from ϱ_1 to ϱ_2 (see Fig. 2.3).

Therefore, given two points $P_1 = (x_1, y_1, z_1)$, $P_2 = (x_2, y_2, z_2)$ in $\mathbb{R}^3$, the horizontal paths which minimizes the length from P_1 to P_2 are the curves $\gamma : [0, 1] \to \mathbb{R}^3$ whose signed area of $\mathscr{D}$ satisfies

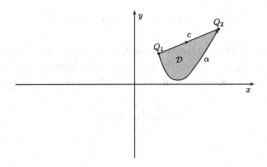

Fig. 2.3 The curve α, the domain $\mathscr{D}$ and the segment c

$$\int_{\mathscr{D}} dx \wedge dy = (z_2 - z_1) - \int_c \frac{1}{2}(xdy - ydx),$$

with minimal length. According to the isoperimetric inequality, the curves in the plane sweeping the same area and which minimize the length are given by circles. This fact can be easily recovered by Theorem 2.9 and Proposition 2.8 (we saw in Example 1.14 that Δ admits no non-trivial singular horizontal paths). Assume that $\gamma_u = (x, y, z) : [0, 1] \to \mathbb{R}^3$ is a minimizing geodesic from $P_1 := \gamma_u(0)$ to $P_2 := \gamma_u(1) \neq P_1$. Then according to Proposition 2.8, there is a smooth arc $p = (p_1, p_2, p_3) : [0, 1] \to (\mathbb{R}^3)^*$ such that the following system of differential equations holds

$$\begin{cases} \dot{x} = p_x - \frac{y}{2}p_z \\ \dot{y} = p_y + \frac{x}{2}p_z \\ \dot{z} = \frac{1}{2}\left((p_y + \frac{x}{2}p_z)x - (p_x - \frac{y}{2}p_z)y\right), \end{cases} \qquad \begin{cases} \dot{p}_x = -\left(p_y + \frac{x}{2}p_z\right)\frac{p_z}{2} \\ \dot{p}_y = \left(p_x - \frac{y}{2}p_z\right)\frac{p_z}{2} \\ \dot{p}_z = 0. \end{cases}$$

Hence $p_z = \bar{p}_z$ for every t. Which implies that

$$\ddot{x} = -\bar{p}_z\dot{y} \quad \text{and} \quad \ddot{y} = \bar{p}_z\dot{x}.$$

If $\bar{p}_z = 0$, then the geodesic from P_1 to P_2 is a segment with constant speed. If $\bar{p}_z \neq 0$, we have or

$$\dddot{x} = -\bar{p}_z^2\dot{x} \quad \text{and} \quad \dddot{y} = -\bar{p}_z^2\dot{y}.$$

Which means that the curve $t \mapsto (x(t), y(t))$ is a circle.

A singular minimizing geodesic. As we said above, minimizing geodesics do not necessarily satisfy the Hamiltonian geodesic equation. As an example, consider the Martinet-like distribution (see Examples 1.10 and 1.16) in $\mathbb{R}^3$ (with coordinates (x_1, x_2, x_3)) generated by

$$X = \partial_{x_1}, \quad Y = (1 + x_1\phi(x))\, \partial_{x_2} + x_1^2\, \partial_{x_3},$$

where ϕ is a smooth function and equipped with a metric g making $\{X, Y\}$ an orthonormal family. In a sufficiently small neighborhood of the origin $\mathcal{V}$, singular curves are given by the horizontal paths which are contained in the Martinet set

$$\Sigma_\Delta = \left\{ x_1 = 0 \right\},$$

that is of the form

$$x(t) = \left(0,\, x_2(0) + \int_0^t u_2(s)\,ds,\, 0,\, x_3(0) \right),$$

with $u_2 \in L^2([0, T]; \mathbb{R})$. Such curves are locally minimizing.

Theorem 2.23 *There is $\bar{\varepsilon} > 0$ such that for every $\varepsilon \in (0, \bar{\varepsilon})$ the horizontal path given by*

$$\bar{\gamma}(t) = (0, t, 0) \quad \forall t \in [0, \varepsilon],$$

minimizes the length among all horizontal paths joining $(0, 0, 0)$ to $(0, \varepsilon, 0)$.

Proof We need to show that among all controls $u = (u_1, u_2) : [0, \tau] \to \mathbb{R}^2$ with $u_1^2 + u_2^2 \le 1$ steering the origin to $P := (0, \varepsilon, 0)$, we have $\varepsilon < \tau$. There is $r > 0$ such that $\bar{B}_{SR}(0, r)$ is included in $\mathcal{V}$. If $\varepsilon \in (0, r)$, then any minimizing geodesic joining 0 to $(0, \varepsilon, 0)$ is contained in $\bar{B}_{SR}(0, r)$. As a matter of fact, we know that $d_{SR}(0, P) \le \varepsilon < r$. Let $C > 0$ be upper bounds for ϕ on $\bar{B}_{SR}(0, r)$ Let $\gamma_u = x : [0, \tau] \to \mathbb{R}^3$ be a competitor for $\bar{\gamma}$. We get easily

$$\begin{cases} x_1(\tau) = \int_0^\tau u_1(s)\,ds = 0 \\ x_2(\tau) = \int_0^\tau u_2(s)\left(1 + x_1(s)\phi(x(s))\right)\,ds = \varepsilon \\ x_3(\tau) = \int_0^\tau u_2(s)x_1(s)^2\,ds = 0. \end{cases} \tag{2.32}$$

Set

$$\beta := \max\left\{ |x_1(s)| \mid s \in [0, \tau] \right\}. \tag{2.33}$$

Note that if $\gamma_u \neq \bar{\gamma}$, then β is necessarily positive. Taking $r > 0$ smaller if necessary (and a fortiori $\varepsilon > 0$ smaller), we may assume that $\beta \le 1/(2C)$. The last equation in (2.32) yields (2.33)

$$\int_0^\tau x_1(s)^2\,ds = \int_0^\tau x_1(s)^2\left(1 - u_2(s)\right)\,ds + \int_0^\tau x_1(s)^2 u_2(s)\,ds$$

$$\le \beta^2 \left(\tau - \int_0^\tau u_2(s)\,ds \right). \tag{2.34}$$

Let $\bar{s} \in [0, \tau]$ be such that $|x_1(\bar{s})| = \beta$. Since $|\dot{x}_1(s)| \le 1$ for almost every $s \in [0, \tau]$ and $x_1(0) = x_1(\tau) = 0$, we have $\bar{s}, \tau - \bar{s} \ge \beta$. Which means that the interval $[\bar{s} - \beta/2, \bar{s} + \beta/2]$ is included in $[0, \tau]$ and

$$|x_1(s)| \ge \frac{\beta}{2} \quad \forall s \in [\bar{s} - \beta/2, \bar{s} + \beta/2].$$

Therefore we have

$$\int_0^\tau x_1(s)^2 \, ds \ge \int_{\bar{s}-\beta/2}^{\bar{s}+\beta/2} x_1(s)^2 \, ds \ge \frac{\beta^3}{4}.$$

By (2.34), we deduce that

$$\frac{\beta^3}{4} \le \beta^2 \left(\tau - \int_0^\tau u_2(s) \, ds \right)$$

which implies

$$\int_0^\tau u_2(s) \, ds \le \tau - \frac{\beta}{4}. \tag{2.35}$$

Then by the second line in (2.32) and the definitions of β and C, we have

$$\varepsilon = \int_0^\tau u_2(s) \, ds + \int_0^\tau u_2(s)x_1(s)\phi(x(s)) \, ds$$
$$\le \int_0^\tau u_2(s) \, ds + \int_0^\tau |u_2(s)| \, |x_1(s)| \, |\phi(x(s))| \, ds$$
$$\le \int_0^\tau u_2(s) \, ds + \beta C \tau.$$

Consequently by (2.35), we get

$$\varepsilon \le \tau + \beta \left(C\tau - \frac{1}{4} \right).$$

In conclusion, if $\beta > 0$ and $\tau < 1/(4C)$ (that is $\bar{\varepsilon} > 0$ small enough), then τ cannot be smaller than ε. This shows the result. ☐

According to Proposition 2.8, for every $\bar{p} = (\bar{p}_1, \bar{p}_2, \bar{p}_3) \in (\mathbb{R}^3)^*$, the normal extremal (with respect to g) on $[0, 1]$ starting at $(0, p)$ is the trajectory $(x, p) : [0, 1] \to \mathbb{R}^3 \times (\mathbb{R}^3)^*$ satisfying

$$\begin{cases} \dot{x}_1 = p_1 \\ \dot{x}_2 = (p \cdot Y(x))(1 + x_1\phi(x)) \\ \dot{x}_3 = (p \cdot Y(x))x_1^2, \end{cases} \tag{2.36}$$

$$\begin{cases} \dot{p}_1 = -\left(p \cdot Y(x)\right)\left[p_2\big(\phi(x) + x_1\frac{\partial \phi}{\partial x_1}(x)\big) + 2p_3 x_1\right] \\ \dot{p}_2 = -\left(p \cdot \hat{Y}(x)\right)\left[p_2 x_1 \frac{\partial \phi}{\partial x_2}(x)\right] \\ \dot{p}_3 = -\left(p \cdot \hat{Y}(x)\right)\left[p_2 x_1 \frac{\partial \phi}{\partial x_3}(x)\right], \end{cases} \qquad (2.37)$$

with

$$p \cdot Y(x) = p_2\big(1 + x_1\phi(x)\big) + p_3 x_1^2$$

and

$$x_1(0) = x_2(0) = x_3(0) = 0, \quad p_1(0) = \bar{p}_1, \quad p_2(0) = \bar{p}_2, \quad p_3(0) = \bar{p}_3. \qquad (2.38)$$

Note that if $\phi \equiv 0$, that is whenever $g = dx_1^2 + dx_2^2$, then the horizontal path given by $\bar{\gamma}(t) = (0, t, 0)$ for any $t \in [0, \varepsilon]$ is the projection of the normal extremal starting at $(0, \bar{p})$ with $\bar{p} = (0, 1, 0)$. Then it is a singular normal minimizing geodesic between its end-points (see Example 1.16 and Theorem 2.9). Different choices of metrics can provide examples of strictly abnormal minimizing geodesics.

Proposition 2.24 *If $\phi(0) \neq 0$, then any reparametrization of $\bar{\gamma}$ is not the projection of a normal extremal.*

Proof We argue by contradiction and assume that there is $\bar{p} = (\bar{p}_1, \bar{p}_2, \bar{p}_3) \in (\mathbb{R}^3)^*$ and $\hat{\gamma} : [0, 1] \to \mathbb{R}^3$ a reparametrization of $\bar{\gamma}$ such that the systems differential Eqs. (2.36) and (2.37) are satisfied with $x = \hat{\gamma}$ and initial conditions (2.38). The system (2.36) and (2.37) is the Hamiltonian system which is associated with the Hamiltonian given by

$$H(x, p) = \frac{1}{2}\,(p \cdot X(x))^2 + \frac{1}{2}\,(p \cdot Y)^2\,.$$

Since H is constant along its extremals and $x_1(t) = x_3(t) = 0$ for any $t \in [0, \varepsilon]$, we have

$$\big(p(t) \cdot X(\bar{\gamma}(t))\big)^2 + \big(p(t) \cdot Y(\bar{\gamma}(t))\big)^2 = p_1(t)^2 + p_2(t)^2 = \bar{p}_1^2 + \bar{p}_2^2 \qquad \forall t \in [0, 1].$$

On the other hand, since $x_1(t) = 0$ for every $t \in [0, 1]$, the second and third equations in (2.37) yield

$$\dot{p}_2 = \dot{p}_3 = 0 \quad \Longrightarrow \quad p_2(t) = \bar{p}_2 \ \forall t \in [0, 1].$$

Moreover, (2.36) also gives $\dot{x}_2 = \bar{p}_2$ that is $\bar{p}_2 \neq 0$ ($\hat{\gamma}$ has constant speed). Since p_1 is smooth and both p_2 and $p_1^2 + p_2^2$ are constant, p_1 is necessarily constant. The first equation in (2.36) and $x_1 = 0$ give $\dot{x}_1 = p_1$. Hence $p_1 = \bar{p}_1 = 0$. Then, using that $p_1 = 0, p_2 = \bar{p}_2 \neq 0$, the first equation in (2.37) gives

$$\bar{p}_2 \phi\big(\hat{\gamma}(t)\big) = 0 \qquad \forall t \in [0, 1].$$

By assumption on $\phi(0)$, we deduce that $\bar{p}_2 = 0$. Since we know that $\hat{\gamma}$ joins 0 to $(0, \varepsilon, 0)$ with $\varepsilon \neq 0$, this contradicts the equality $\dot{x}_2 = \bar{p}_2$. $\qquad \square$

2.6 Notes and Comments

Theorem 2.9 may be seen as a weak form of the Pontryagin maximum principle which has been developed by the russian school of control in the 60s. In the general context of optimal control theory, the strong form of the Pontryagin maximum principle provides necessary conditions for a control to be optimal. For further details on this topics, we refer the reader to the seminal book by Pontryagin and its collaborators [9] and to the more recent textbooks by Agrachev and Sachkov [2], Clarke [4], or Vinter [11]. The material presented in Sects. 2.1 and 2.2 is by now classical. It can be found in the Montgomery textbook [8] which also provides many references.

Theorem 2.14 about the image of the sub-Riemannian exponential map has been proven by Agrachev and the author, see [1]. It extends a previous density result, based on Lemma 2.15, which was obtained by Trélat and the author in [10]. Given a complete sub-Riemannian structure (Δ, g) on a smooth manifold M and $x \in M$, we do not know if the image of $\exp_x$ has full Lebesgue measure in M. This open problem is indeed "contained" in the sub-Riemannian Sard conjecture. Given $x \in M$ (which is equipped with a SR structure), denote by $\mathscr{S}_\Delta^{x,1}$ the set of singular horizontal paths in $\Omega_\Delta^{x,1}$ (that is $\mathscr{S}_\Delta^{x,1} := \Omega_\Delta^{x,1} \setminus \mathscr{R}_\Delta^{x,1}$ with the notations of Chap. 1). The SR Sard conjecture states that the image of $\mathscr{S}_\Delta^{x,1}$ by the End-Point mapping

$$E_\Delta^{x,T} : \Omega_\Delta^{x,1} \longrightarrow M$$
$$\gamma \longmapsto \gamma(1),$$

has Lebesgue measure zero in M. We even do not know if $E_\Delta^{x,T}$ can have a non-empty interior in M. We refer the reader to Montgomery's book [8] for further details on the SR Sard Conjecture and to the paper [10] for various sub-Riemannian Sard-like conjectures.

The theory of second variation for singular geodesics in sub-Riemannian geometry has been developed by Agrachev and its collaborators. The results and proofs that we present in Sect. 2.4 are taken from Agrachev-Sarychev's paper [3]. Example 2.1 (medium-fat distributions) is taken from [3] as well.

For decades the prevailing wisdom was that every sub-Riemannian minimizing geodesic is normal, meaning that it admits a normal extremal lift. In 1991, Montgomery [7] found the first counter-example to this assertion. We refer the reader to Montgomery's book [8] for an historical account on the existence of strictly abnormal minimizing geodesics. The second example which is presented in Sect. 2.5 is moreorless the Montgomery counter-example. The proof of local minimality of characteristic lines in the Martinet surface (Theorem 2.23) is taken from the monograph by Liu and Sussmann [6] which provide a more general class of counter-examples. Note that the Montgomery counter-example as well as all other known counter-examples exhibit smooth singular minimizing curves. The existence of non-smooth sub-Riemannian geodesics is open.

In the first example of Sect. 2.5, we briefly explained that the sub-Riemannian structure under study was indeed left-invariant under some group law. This additional structure makes $\mathbb{H}^1$ a Carnot group. We refer the reader to the Montgomery textbook [8] or to the Jean monograph [5] for further details on Carnot groups.

References

1. Agrachev, A.: Any sub-Riemannian metric has points of smoothness. Dokl. Akad. Nauk. 424(3), 295–298 (2009), translation in. Dokl. Math. **79**(1), 45–47 (2009)
2. Agrachev, A.A., Sachkov, Y.L.: Control Theory from the Geometric Viewpoint. Encyclopaedia of Mathematical Sciences, vol. 87, Springer, Heidelberg (2004) NULL
3. Rifford, L., Trélat, E.: Morse-Sard type results in sub-riemannian geometry. Math. Ann. **332**(1), 145–159 (2005)
4. Clarke, F.H.: Optimization and Nonsmooth Analysis. Wiley-Interscience, New York (1983) Republished as vol. 5 of Classics in Applied Mathematics, SIAM (1990)
5. Jean, F.: Control of Nonholonomic Systems and Sub-Riemannian geometry. Lectures given at the CIMPA School "Géométrie sous-riemannienne", Beirut (2012)
6. Montgomery, R.: Abnormal minimizers. SIAM J. Control Optim. **32**(6), 1605–1620 (1994)
7. Agrachev, A., Sarychev, A.: Sub-Riemannian metrics: minimality of singular geodesics versus subanalyticity. ESAIM Control Optim. Calc. Var. **4**, 377–403 (1999)
8. Montgomery, R.: A tour of subriemannian geometries, their geodesics and applications. Mathematical Surveys and Monographs, vol. 91. American Mathematical Society, Providence, RI (2002)
9. Pontryagin, L., Boltyanskii, V., Gamkrelidze, R., Mischenko, E.: The Mathematical Theory of Optimal Processes. Wiley-Interscience, New-York (1962)
10. Liu, W., Sussmann, H.J.: Shortest paths for sub-Riemannian metrics on rank-2 distributions. Mem. Amer. Math. Soc. **118**, 564 (1995)
11. Vinter, R.B.: Optimal Control. Birkhäuser, Boston (2000)

Chapter 3
Introduction to Optimal Transport

Abstract This Chapter is concerned with the study of optimal transport maps in the sub-Riemannian setting. We first provide a course in optimal transport theory. Then we study the well-posedness of the Monge problem for sub-Riemannian quadratic costs.

Throughout all the chapter, M denotes a smooth connected manifold without boundary of dimension $n \geq 2$.

3.1 The Monge and Kantorovitch Problems

The Monge problem. Let

$$c : M \times M \to [0, +\infty)$$

be a *cost function* and μ, ν be two probability measures on M. We recall that a probability measure on M is a Borel measure with total mass 1. The *Monge optimal transport problem* from μ to ν with respect to the cost c consists in minimizing the transportation cost

$$\int_M c(x, T(x)) \, d\mu(x),$$

among all the measurable maps $T : M \to M$ pushing forward μ to ν (we denote it by $T_\sharp \mu = \nu$) that is satisfying

$$\mu(T^{-1}(B)) = \nu(B) \qquad \forall B \text{ measurable set in } M.$$

Such maps are called *transport maps* from μ to ν (Fig. 3.1).
    We set

L. Rifford, *Sub-Riemannian Geometry and Optimal Transport*,
SpringerBriefs in Mathematics, DOI: 10.1007/978-3-319-04804-8_3,
© The Author(s) 2014

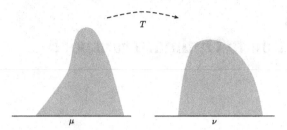

Fig. 3.1 The Monge problem

$$C_{\mathcal{M}}(\mu, \nu) := \inf \left\{ \int_M c\big(x, T(x)\big) \, d\mu \mid T_{\sharp}\mu = \nu \right\}, \tag{3.1}$$

where $T_{\sharp}\mu = \nu$ means implicitly that T is a measurable map from M to itself which pushes forward μ to ν.

Remark 3.1 The property (3.1) is equivalent to

$$\int_M \varphi(T(x)) \, d\mu(x) = \int_M \varphi(y) \, d\nu(y),$$

for all ν-integrable function φ. If $M = \mathbb{R}^n$ and μ and ν are absolutely continuous with respect to the Lebesgue measure respectively with densities f and g in $L^1(\mathbb{R}^n; [0, +\infty))$, the latter property can be written as

$$\int_M \varphi(T(x)) f(x) \, dx = \int_M \varphi(y) g(y) \, dy,$$

for any $\varphi \in L^\infty(\mathbb{R}^n; \mathbb{R})$. Therefore, if T is a diffeomorphism, then the change of variable $y = T(x)$ yields the Monge-Ampère equation

$$\left| \det\big(D_x T\big) \right| = \frac{f(x)}{g(T(x))} \qquad \mu - \text{a.e. } x \in \mathbb{R}^n.$$

Example 3.1 Transport maps may not exist. For example, consider in $\mathbb{R}^n$ the probability measures μ, ν given by

$$\mu = \delta_x \quad \text{and} \quad \nu = \frac{1}{2}\delta_{y_1} + \frac{1}{2}\delta_{y_2},$$

where $x, y_1, y_2 \in \mathbb{R}^n$, $y_1 \neq y_2$ and δ_a denotes the Dirac mass at some point $a \in \mathbb{R}^n$. There are no transport maps from μ to ν. If such a map T exists, then

$$\frac{1}{2} = \nu(\{y_1\}) = \mu\left(T^{-1}(\{y_1\})\right) = 0 \text{ or } 1,$$

which is impossible.

Example 3.2 Minimizers of Monge's problem may not be unique. On the real line $\mathbb{R}$, consider the probability measures μ and ν given by

$$\mu = 1_{[0,\,1]}\,\mathscr{L}^1 \quad \text{and} \quad \nu = 1_{[1,\,2]}\,\mathscr{L}^1,$$

where $\mathscr{L}^1$ denotes the Lebesgue measure in $\mathbb{R}$. In other terms, μ and ν are respectively the restriction of the Lebesgue measure on the intervals $[0, 1]$ and $[1, 2]$. The two maps $T_1, T_2 : \mathbb{R} \to \mathbb{R}$ given by

$$T_1(x) = x + 1 \quad \text{and} \quad T_2(x) = 2 - x \qquad \forall x \in \mathbb{R},$$

push forward μ to ν. This is a straightforward consequence of the fact that both T_1 and T_2 are affine maps which are bijective from $[0, 1]$ to $[1, 2]$ with determinant 1 together with a change of variable (see Remark 3.1). Consider the Monge cost $c : \mathbb{R} \times \mathbb{R} \to [0, +\infty)$ given by

$$c(x, y) := |y - x| \qquad \forall x, y \in \mathbb{R}.$$

We check easily that the transportation cost for T_1 and T_2 are given by

$$\int_{\mathbb{R}} c\big(x, T_i(x)\big)\,d\mu(x) = \int_0^1 |T_i(x) - x|\,dx = 1 \qquad i = 1, 2.$$

Furthermore, we also check that if T is a map which pushes forward μ to ν, then

$$\begin{aligned}
\int_{\mathbb{R}} c\big(x, T(x)\big)\,d\mu(x) &= \int_0^1 |T(x) - x|\,dx \\
&= \int_0^1 [T(x) - x]\,dx = \int_0^1 T(x)\,dx - \int_0^1 x\,dx \\
&= \int_1^2 y\,dy - \int_0^1 x\,dx = 1.
\end{aligned}$$

This shows that the infimum in the definition of $C_{\mathscr{M}}(\mu, \nu)$ is attained by all transport maps from μ to ν. So, it is not unique.

The constraint $T_\sharp \mu = \nu$ being highly non-linear, the Monge optimal transport problem is quite difficult from the viewpoint of optimization. That is why we will study a notion of weak solution for this problem.

The Kantorovitch relaxation. Given two probability measures μ, ν on M, we denote by $\Pi(\mu, \nu)$ the set of probability measures α in the product $M \times M$ with first and second marginals μ and ν, that is such that

$$\pi_\sharp^1 \alpha = \mu \quad \text{and} \quad \pi_\sharp^2 \alpha = \nu, \tag{3.2}$$

where $\pi^i : M \times M \to M$ denotes respectively the projection on the first and second variable in $M \times M$. The *Kantorovitch optimal transport problem* with respect to the cost $c : M \times M \to [0, +\infty)$ consists in minimizing the quantity

$$C(\alpha) := \int_{M \times M} c(x, y) \, d\alpha(x, y),$$

among all the $\alpha \in \Pi(\mu, \nu)$. Any measure in $\alpha \in \Pi(\mu, \nu)$ is called a *transport plan* between μ and ν. We set

$$C_{\mathscr{K}}(\mu, \nu) := \inf \left\{ C(\alpha) \mid \alpha \in \Gamma(\mu, \nu) \right\}.$$

Remark 3.2 The property (3.2) is equivalent to

$$\mu(B) = \alpha(B \times M) \quad \text{and} \quad \nu(B) = \alpha(M \times B),$$

for any measurable set B in M, which is also equivalent to

$$\int_{M \times M} \left[\varphi_1(x) + \varphi_2(y) \right] d\alpha(x, y) = \int_M \varphi_1(x) \, d\mu(x) + \int_M \varphi_2(y) \, d\nu(y),$$

for all μ-integrable function φ_1 and ν-integrable function φ_2. In particular, the set $\Pi(\mu, \nu)$ is a convex set which always contains the product measure $\mu \times \nu$.

Remark 3.3 If $T : M \to M$ is a transport map from μ to ν then the measure α on $M \times M$ given by

$$\alpha := (Id \times T)_\sharp \, \mu,$$

is a transport plan between μ and ν. This means that the Kantorovitch optimization problem is more general than the Monge optimization problem, or

$$C_{\mathscr{K}}(\mu, \nu) \le C_{\mathscr{M}}(\mu, \nu),$$

for all probability measures μ, ν on M.

Example 3.3 Returning to Example 3.1, we note that the product measure

$$\alpha = \frac{1}{2} \delta_{(x, y_1)} + \frac{1}{2} \delta_{(x, y_2)},$$

is a transport plan between μ and ν. In contrary to Monge's transport maps, Kantorovitch's transport plans allow splitting of mass.

The Kantorovitch optimal transport problem is an infinite-dimensional optimization problem which involves a functional C which is linear in α and a set of constraints $\Pi(\mu, \nu)$ which is convex and weakly compact. The existence of optimal transport plans becomes easy.

3.2 Optimal Plans and Kantorovitch Potentials

Optimal plans. Throughout this section, we fix a cost $c : M \times M \to [0, +\infty)$. We recall that the *support* $\mathrm{spt}(\mu)$ of a measure μ refers to the smallest closed set $F \subset M$ of full mass $\mu(F) = \mu(M) = 1$.

Theorem 3.1 *Let μ, v be two probability measures on M. Assume that c is continuous and that $\mathrm{Supp}(\mu)$ and $\mathrm{Supp}(v)$ are compact. Then the Kantorovitch optimal transport problem admits at least one solution, that is there is $\bar{\alpha} \in \Pi(\mu, v)$ such that*

$$C(\bar{\alpha}) = C_{\mathscr{K}}(\mu, v) := \inf \Big\{ C(\alpha) \,|\, \alpha \in \Gamma(\mu, v) \Big\}.$$

Proof We first note that $C_{\mathscr{K}}(\mu, v)$ is finite. As a matter of fact, since the product measure $\mu \times v$ belongs to $\Pi(\mu, v)$ and c is bounded on $\mathrm{Supp}(\mu) \times \mathrm{Supp}(v)$ (by assumption c is continuous and $\mathrm{Supp}(\mu), \mathrm{Supp}(v)$ are compact), we have $C_{\mathscr{K}}(\mu, v) \leq C(\mu \times v) < +\infty$. In fact, the supports of all transport plans between μ and v are contained in the set $\mathrm{Supp}(\mu) \times \mathrm{Supp}(v) \subset M \times M$ which is compact by assumption on $\mathrm{Supp}(\mu)$ and $\mathrm{Supp}(v)$. Then we can assume without loss of generality that M is compact. Denote by $\mathscr{P}(M \times M)$ the set of probability measures on $M \times M$ and define $F : \mathscr{P}(M \times M) \to \mathbb{R}$ by

$$F(\alpha) := \int_{M \times M} c(x, y) \, d\alpha(x, y) \qquad \forall \alpha \in \mathscr{P}(M \times M).$$

The functional F is continuous on $\mathscr{P}(M \times M)$ equipped with the topology of weak convergence, that is for any sequence $\{\alpha_k\}_k$ and any α in $\mathscr{P}(M \times M)$ satisfying

$$\int_{M \times M} \varphi(x, y) \, d\alpha_k(x, y) \longrightarrow_{k \to +\infty} \int_{M \times M} \varphi(x, y) \, d\alpha(x, y),$$

for any measurable function $\varphi : M \to \mathbb{R}$ which is bounded, we have

$$\lim_{k \to +\infty} F(\alpha_k) = F(\alpha).$$

This fact is a straigthforward consequence of the continuity of c together with the compactness of $M \times M$. By Prokhorov's Theorem, the set of probability measures on $M \times M$ is compact with respect to weak convergence. We conclude easily. Let $\{\alpha_k\}_k$ be a sequence in $\Pi(\mu, v)$ such that

$$C_{\mathscr{K}}(\mu, v) = \lim_{k \to +\infty} C(\alpha_k).$$

By Prokhorov's Theorem, up to taking a subsequence, we may assume that $\{\alpha_k\}$ converges to some probability measure $\bar{\alpha}$. By Remark 3.2, $\bar{\alpha}$ belongs to $\Pi(\mu, v)$. Moreover it satisfies $C(\bar{\alpha}) = C_{\mathscr{K}}(\mu, v)$ by continuity of F. $\qquad\square$

The supports of optimal transport plans have specific properties. Let us introduce the notion of c-cyclically monotone sets.

Definition 3.2 A subset $S \subset M \times M$ is called *c-cyclically monotone* if for any finite number of points $(x_j, y_j) \in S, j = 1, \ldots, J$, and σ a permutation on the set $\{1, \ldots, J\}$,

$$\sum_{j=1}^{J} c(x_j, y_j) \leq \sum_{j=1}^{J} c\left(x_{\sigma(j)}, y_j\right).$$

Remark 3.4 The definition given above is equivalent to the following one: for any finite number of points $(x_j, y_j) \in S, j = 1, \ldots, J$,

$$\sum_{j=1}^{J} c(x_j, y_j) \leq \sum_{j=1}^{J} c\left(x_j, y_{j+1}\right),$$

with $y_{J+1} = y_1$. The equivalence is a straightforward consequence of the decomposition of a permutation into disjoint commuting cycles.

Remark 3.5 If c is assumed to be continuous, the c-cyclical monotonocity is stable under closure. The closure of a c-cyclically monotone set is c-cyclically monotone.

Given two probability measures μ, ν on M, we call *optimal transport plan* between μ and ν any $\alpha \in \Pi(\mu, \nu)$ satisfying $C_{\mathscr{H}}(\mu, \nu) = C(\alpha)$. Optimal transport plans always have c-cyclically monotone supports.

Theorem 3.3 *Let μ, ν be two probability measures on M. Assume that c is continuous and that $Supp(\mu)$ and $Supp(\nu)$ are compact. Then there is a c-cyclically monotone compact set $\mathscr{S} \subset Supp(\mu) \times Supp(\nu)$ such that the support of any optimal transport plan between μ and ν is contained in $\mathscr{S}$.*

Proof Let us first show that the supports of optimal transport plans are always c-cyclically monotone. We argue by contradiction and assume that there is an optimal transport plan $\alpha \in \Pi(\mu, \nu)$ whose support is not c-cyclically monotone. Then there is an integer $J > 1$, J points $(x_1, y_1), \ldots, (x_J, y_J)$ in $Supp(\alpha)$ and a permutation σ on the set $\{1, \ldots, J\}$ such that

$$\sum_{j=1}^{J} c(x_j, y_j) > \sum_{j=1}^{J} c\left(x_{\sigma(j)}, y_j\right).$$

By continuity of c, there are open sets U_j, V_j for $j = 1, \ldots, J$ which contain respectively x_j, y_j such that

$$\sum_{j=1}^{J} c(u_j, v_j) > \sum_{j=1}^{J} c\left(u_{\sigma(j)}, v_j\right) \qquad \forall \left((u_j, v_j)\right)_{j=1,\ldots,J} \in \Pi_{j=1}^{J}(U_j \times V_j).$$

$$(3.3)$$

Each (x_j, y_j) belongs to the support of α, then we have $\alpha(U_j \times V_j) > 0$. Define the probability measure P on $\Pi_{j=1}^{J}(U_j \times V_j)$ by

$$P = \Pi_{j=1}^{J}\left[\frac{1}{\alpha\left(U_j \times V_j\right)} 1_{U_j \times V_j}\alpha\right].$$

It is a product of probability measures, hence it is a probability measure as well. Set

$$\bar{m} := \min\left\{\alpha(U_j \times V_j) \mid j = 1, \ldots, J\right\},$$

denote by π^{U_j} (resp. π^{V_j}) the projection from $\Pi_{j=1}^{J}(U_j \times V_j)$ to U_j (resp. to V_j) and define the measure $\tilde{\alpha}$ on $M \times M$ by

$$\tilde{\alpha} = \alpha + \frac{\bar{m}}{J}\left[\sum_{j=1}^{J}\left(\left(\pi^{U_{\sigma(j)}}, \pi^{V_j}\right)_{\sharp}P - \left(\pi^{U_j}, \pi^{V_j}\right)_{\sharp}P\right)\right].$$

We have

$$\tilde{\alpha} \geq \alpha - \frac{\bar{m}}{J}\sum_{j=1}^{J}\left(\pi^{U_j}, \pi^{V_j}\right)_{\sharp}P = \alpha - \frac{1}{J}\sum_{j=1}^{J}\frac{\bar{m}}{\alpha\left(U_j \times V_j\right)}1_{U_j \times V_j}\alpha$$

$$\geq \alpha - \frac{1}{J}\sum_{j=1}^{J}1_{U_j \times V_j}\alpha \geq \alpha - \alpha = 0.$$

Moreover

$$\pi_{\sharp}^{1}\left(\sum_{j=1}^{J}\left(\pi^{U_j}, \pi^{V_j}\right)_{\sharp}P\right) = \sum_{j=1}^{J}\pi_{\sharp}^{U_j}P = \sum_{j=1}^{J}\pi_{\sharp}^{U_{\sigma(j)}}P = \pi_{\sharp}^{1}\left(\sum_{j=1}^{J}\left(\pi^{U_{\sigma(j)}}, \pi^{V_j}\right)_{\sharp}P\right)$$

and

$$\pi_{\sharp}^{2}\left(\sum_{j=1}^{J}\left(\pi^{U_j}, \pi^{V_j}\right)_{\sharp}P\right) = \sum_{j=1}^{J}\pi_{\sharp}^{V_j}P = \pi_{\sharp}^{2}\left(\sum_{j=1}^{J}\left(\pi^{U_{\sigma(j)}}, \pi^{V_j}\right)_{\sharp}P\right).$$

Therefore α is a non-negative measure which belongs to $\Pi(\mu, \nu)$. But by construction, we have

$$\int_{M \times M}c(x, y)\,d\tilde{\alpha}(x, y) = \int_{M \times M}c(x, y)\,d\alpha(x, y)$$

$$+ \frac{\bar{m}}{J}\int\sum_{j=1}^{J}\left(c\left(u_{\sigma(j)}, v_j\right) - c\left(u_j, v_j\right)\right)dP\left((u_1, v_1,), \ldots, (u_J, v_J)\right),$$

and the last term is negative [by (3.3)]. This means that α cannot be optimal and gives a contradiction. Then we know that the supports of any optimal transport plan between μ and ν is c-cyclically monotone. Denote by $\Pi^{opt}(\mu, \nu)$ the set of optimal transport plans in $\Pi(\mu, \nu)$ and set

$$\mathscr{S} := \bigcup_{\alpha \in \Pi^{opt}(\mu,\nu)} \mathrm{Supp}(\alpha).$$

By construction, $\mathscr{S}$ is a subset of $\mathrm{Supp}(\mu) \times \mathrm{Supp}(\nu) \subset M \times M$ which contains the supports of all optimal transport plans. It remains to show that $\mathscr{S}$ is c-cyclically monotone. Let $(x_1, y_1), \ldots, (x_J, y_J)$ be J points in $\mathscr{S}$ and σ be a permutation on the set $\{1, \ldots, J\}$. For each $j = 1, \ldots, J$ the point (x_j, y_j) belongs to the support of an optimal transport plan α_j. Let $\bar{\alpha}$ be the convex combination of the α_j's, that is

$$\alpha := \frac{1}{J} \sum_{j=1}^{J} \alpha_j.$$

Since $\Pi(\mu, \nu)$ is convex and the mapping $\alpha \mapsto C(\alpha)$ is linear, $\bar{\alpha}$ belongs to $\Pi^{opt}(\mu, \nu)$. Then its support is c-cyclically monotone and contains all the (x_j, y_j)'s. We infer that

$$\sum_{j=1}^{J} c(x_j, y_j) \le \sum_{j=1}^{J} c\left(x_{\sigma(j)}, y_j\right).$$

We conclude by Remark 3.5.   □

Example 3.4 Returning to Example 3.2, we can show that the set provided by Theorem 3.3 has to be $\mathscr{S} = [0, 1] \times [1, 2] = \mathrm{Supp}(\mu) \times \mathrm{Supp}(\nu)$. As a matter of fact, for every $(x, y) \in [0, 1] \times [1, 2]$ there is a bijective function $T : [0, 1] \to [1, 2]$ which is lower semicontinuous, increasing and piecewise affine with slope 1, and whose graph contains (x, y) (see Fig. 3.2). Thanks to the observation we did in Example 3.2, such a function is a transport map from $\mu = 1_{[0, 1]}\mathscr{L}^1$ to $\nu = 1_{[1, 2]}\mathscr{L}^1$, hence it is optimal.

Kantorovitch potentials. The aim of this section is to characterize c-cyclically monotone sets in a more analytic way.

Definition 3.4 A function $\psi : M \to \mathbb{R} \cup \{+\infty\}$, not identically $+\infty$, is said to be *c-convex* if there is a non-empty set $\mathscr{A} \subset M \times \mathbb{R}$ such that

$$\psi(x) := \sup\left\{\lambda - c(x, y) \mid (y, \lambda) \in \mathscr{A}\right\} \qquad \forall x \in M. \tag{3.4}$$

The *c-transform* of ψ, denoted by ψ^c is the function $\psi^c : M \to \mathbb{R} \cup \{-\infty\}$ defined by

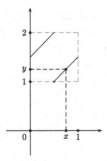

Fig. 3.2 Solution to Example 3.4

$$\psi^c(y) := \inf\left\{\psi(x) + c(x, y) \,|\, x \in M\right\} \qquad \forall y \in M.$$

The pair (ψ, ψ^c) is called a c-pair of *Kantorovitch potentials*.

The following result shows that the opposite of a c-convex function is the c-transform of the opposite of its c-transform.

Proposition 3.5 *Given a c-convex function ψ, the function $-\psi^c$ is c-convex and we have*

$$\psi(x) = \sup\left\{\psi^c(y) - c(x, y) \,|\, y \in M\right\} \qquad \forall x \in M.$$

Proof By definition of ψ^c we have

$$\psi^c(y) - c(x, y) \leq \psi(x) \qquad \forall x \in M, \forall y \in M.$$

Which implies that $\psi(x) \geq \sup_{y \in M}\{\psi^c(y) - c(x, y)\}$ for any $x \in M$. Let us show that $\psi(x) \leq \sup_{y \in M}\{\psi^c(y) - c(x, y)\}$ for any $x \in M$. Argue by contradiction and assume that there is $\bar{x} \in M$ such that

$$\psi(\bar{x}) > \sup\left\{\psi^c(y) - c(\bar{x}, y)\, y \in M\right\}.$$

Since ψ is c-convex, there are a set $\mathscr{A} \subset M \times \mathbb{R}$, $(\bar{y}, \bar{\lambda}) \in \mathscr{A}$ and $\delta > 0$ such that

$$\bar{\lambda} - c(\bar{x}, \bar{y}) + \delta \geq \psi(\bar{x}) \geq \sup\left\{\psi^c(y) - c(\bar{x}, y) \,|\, y \in M\right\} + 3\delta.$$

Then we get $\psi^c(\bar{y}) \leq \bar{\lambda} - 2\delta$, which by definition of $\psi^c(\bar{y})$ implies that there is $x \in M$ such that

$$\psi(x) + c(x, \bar{y}) \leq \bar{\lambda} - \delta.$$

This contradicts (3.4). $\qquad\qquad\square$

Example 3.5 If $M = \mathbb{R}^n$ and c is given by $c(x, y) = |y - x|$, then the c-convex functions are exactly the functions which are 1-Lipschitz on $\mathbb{R}^n$. As a matter of fact, if $f : \mathbb{R}^n \to \mathbb{R}$ is 1-Lipschitz then for every $x \in \mathbb{R}^n$,

$$f(x) \geq f(y) - |y - x| \quad \forall y \in \mathbb{R}^n,$$

which yields

$$f(x) = \sup\Big\{f(y) - c(x, y) \,|\, y \in \mathbb{R}^n\Big\}.$$

Moreover, f is its own c-transform. Conversely, any c-convex function is a supremum of 1-Lipschitz function which is not identically $+\infty$. Then it is finite everywhere and 1-Lipschitz.

Example 3.6 If $M = \mathbb{R}^n$ and c is given by $c(x, y) = |y - x|^2/2$, then the c-convex functions are the functions $\psi : \mathbb{R}^n \to \mathbb{R} \cup \{+\infty\}$ such that the function

$$x \in \mathbb{R}^n \longmapsto \psi(x) + \frac{1}{2}|x|^2$$

is convex. As a matter of fact, any c-convex function can be written as

$$\psi(x) = \sup\Big\{\lambda - \frac{|y|^2}{2} - \langle x, y \rangle \,|\, (y, \lambda) \in \mathscr{A}\Big\} - \frac{|x|^2}{2} \quad \forall x \in \mathbb{R}^n.$$

which shows that $\psi + |\cdot|^2/2$ is convex as a supremum of affine functions. Conversely, any convex function on $\mathbb{R}^n$ can be expressed as the supremum of affine functions. That is given a convex function $\varphi : \mathbb{R}^n \to \mathbb{R} \cup \{+\infty\}$, there is a set $\mathscr{B} \subset \mathbb{R}^n \times \mathbb{R}$ such that

$$\varphi(x) = \sup\Big\{\langle x, y \rangle + \beta \,|\, (y, \beta) \in \mathscr{B}\Big\} \quad \forall x \in \mathbb{R}^n.$$

Then for every $x \in \mathbb{R}^n$,

$$\varphi(x) - \frac{1}{2}|x|^2 = \sup\left\{\Big(\beta + \frac{|y|^2}{2}\Big) - \frac{|y - x|^2}{2} \,|\, (y, \beta) \in \mathscr{B}\right\},$$

which shows that $\psi := \varphi - |\cdot|^2/2$ is c-convex.

The c-cyclically monotone sets are the sets which are contained in the c-subdifferential of c-convex functions.

Definition 3.6 Let $\psi : M \to \mathbb{R} \cup \{+\infty\}$ be a c-convex function. For every $x \in M$, the *c-subdifferential* of ψ at x is defined by

$$\partial_c \psi(x) := \Big\{y \in M \,|\, \psi^c(y) = \psi(x) + c(x, y)\Big\}.$$

We call *contact set* of the pair (ψ, ψ^c) the set defined by

$$\partial_c \psi := \left\{ (x, y) \in M \times M \mid y \in \partial_c \psi(x) \right\}.$$

Remark 3.6 By the above definitions, a pair (x, y) in $M \times M$ belongs to $\partial_c \psi$ if and only if

$$\psi(x) + c(x, y) \leq \psi(z) + c(z, y) \qquad \forall z \in M,$$

which is also equivalent to

$$\psi^c(y) - c(x, y) \geq \psi^c(z) - c(x, z) \qquad \forall z \in M.$$

In particular, both $\psi(x)$ and $\psi^c(y)$ are finite.

The following result is the cornerstone of the results of existence and uniqueness of optimal transport maps that we will present in the next sections.

Theorem 3.7 *For $S \subset M \times M$ to be c-cyclically monotone, it is necessary and sufficient that $S \subset \partial_c \psi$ for some c-convex $\psi : M \to \mathbb{R} \cup \{+\infty\}$. In fact, for every c-cyclically monotone set $S \subset M \times M$, there is a c-pair of potentials (ψ, ψ^c) with $S \subset \partial_c \psi$ satisfying*

$$\psi(x) = \sup\left\{ \psi^c(y) - c(x, y) \mid y \in \pi^2(S) \right\} \qquad \forall x \in M, \tag{3.5}$$

$$\psi^c(y) = \inf\left\{ \psi(x) + c(x, y) \mid x \in \pi^1(S) \right\} \qquad \forall y \in M. \tag{3.6}$$

If c is continuous and S is compact, then both ψ, ψ^c are valued in $\mathbb{R}$ and continuous, and the infimum and supremum in (3.5)–(3.6) are attained.

Proof First, given a c convex function $\psi : X \to \mathbb{R} \cup \{+\infty\}$ the contact set of (ψ, ψ^c) is c-cyclically monotone. As a matter of fact, given $(x_j, y_j) \in \partial_c \psi, j = 1, \ldots, J$, and σ a permutation on the set $\{1, \ldots, J\}$, we have

$$\psi^c(y_j) = \psi(x_j) + c(x_j, y_j) \quad \text{and} \quad \psi^c(y_j) \leq \psi(x_{\sigma(j)}) + c(x_{\sigma(j)}, y_j),$$

for every $j = 1, \ldots, J$. Hence

$$\sum_{j=1}^{J} c(x_j, y_j) = \sum_{j=1}^{J} \psi^c(y_j) - \sum_{j=1}^{J} \psi(x_j) = \sum_{j=1}^{J} \psi^c(y_j) - \sum_{j=1}^{J} \psi(x_{\sigma(j)})$$

$$\leq \sum_{j=1}^{J} c(x_{\sigma(j)}, y_j).$$

Let us now show that a c-cyclically monotone set $S \subset M \times M$ is necessarily included in the contact set of some c-convex function. Fix $(\bar{x}, \bar{y})$ in the c-cyclically monotone set $S \subset M \times M$ and define $\psi : M \to \mathbb{R} \cup \{+\infty\}$ by

$$\psi(x) := \sup \Big\{ \big[c(\bar{x}, \bar{y}) - c(x_1, \bar{y}) \big]$$
$$+ \sum_{j=1}^{J-1} \big[c(x_j, y_j) - c(x_{j+1}, y_j) \big] + \big[c(x_J, y_J) - c(x, y_J) \big]$$
$$| J \in \mathbb{N}, J \geq 2, (x_j, y_j) \in S, \forall j = 1, \dots, J \Big\},$$

for every $x \in M$. We claim that ψ is a c-convex function whose contact set contains S. First taking $J = 2$, $x = x_1 = x_2 = \bar{x}$ and $y_1, y_2 = \bar{y}$, we check easily that $\psi(\bar{x}) \geq 0$. Furthermore, by c-cyclical monotonicity of S, we have

$$\big[c(\bar{x}, \bar{y}) - c(x_1, \bar{y}) \big] + \sum_{j=1}^{J-1} \big[c(x_j, y_j) - c(x_{j+1}, y_j) \big] + \big[c(x_J, y_J) - c(\bar{x}, y_J) \big] \leq 0,$$

for any pairs $(x_1, y_1), \dots, (x_J, y_J)$ belonging to S. Thus we have $\psi(\bar{x}) \leq 0$ and in turn $\psi(\bar{x}) = 0$. This shows that ψ is not identically $+\infty$. Define $\phi : M \to \mathbb{R} \cup \{-\infty\}$ by

$$\phi(y) := \sup \Big\{ \big[c(\bar{x}, \bar{y}) - c(x_1, \bar{y}) \big] + \sum_{j=1}^{J-1} \big[c(x_j, y_j) - c(x_{j+1}, y_j) \big] + c(x_J, y)$$
$$| J \in \mathbb{N}, J \geq 2, (x_j, y_j) \in S, \forall j = 1, \dots, J-1, (x_J, y) \in S \Big\} \quad \forall y \in M.$$

Note that if $y \in M$ is such that there are no $x \in M$ with $(x, y) \in S$, then $\phi(y) = -\infty$. However, as above we check easily that $\phi(\bar{y}) = 0$ which shows that ϕ is not identically $-\infty$. Therefore, by construction we have for every $x \in M$,

$$\psi(x) = \sup \Big\{ \phi(y) - c(x, y) \mid y \in \pi^2(S) \Big\} = \sup \Big\{ \phi(y) - c(x, y) \mid y \in M \Big\}, \quad (3.7)$$

which shows that ψ is c-convex. It remains to check that $S \subset \partial_c \psi$. Let $(x, y) \in S$ be fixed, we need to show that

$$\psi(x) + c(x, y) \leq \psi(z) + c(z, y) \quad \forall z \in M.$$

By construction of ψ, we have for every $z \in M$,

$$\psi(z) \geq \sup \Big\{ \big[c(\bar{x}, \bar{y}) - c(x_1, \bar{y}) \big] + \sum_{j=1}^{J-1} \big[c(x_j, y_j) - c(x_{j+1}, y_j) \big]$$
$$+ \big[c(x, y) - c(z, y) \big] \mid J \in \mathbb{N}, J \geq 2, (x_j, y_j) \in S,$$

$$\forall j = 1, \ldots, J - 1, x_J = x \Big\}$$
$$= \psi(x) + c(x, y) - c(z, y).$$

We get the necessary and sufficient condition for a set to be c-cyclically monotone. Let us now turn to the second part of the result, that is let us prove that for any c-cyclically monotone set $S \subset M \times M$, there is a c-pair of potentials ψ, ψ^c with $S \subset \partial_c \psi$ which in addition satisfies (3.5)–(3.6).

Let S be a c-cyclically monotone set. We already know that there is $\phi : M \to \mathbb{R} \cup \{-\infty\}$ which is not identically $-\infty$ such that the function $\psi : M \to \mathbb{R} \cup \{+\infty\}$ defined by

$$\psi(x) := \sup \Big\{ \phi(y) - c(x, y) \mid y \in \pi^2(S) \Big\} \qquad \forall x \in M, \tag{3.8}$$

is c-convex with $S \subset \partial_c \psi$ (remember (3.7)). Let $\phi_1 = \psi^c : M \to \mathbb{R} \cup \{-\infty\}$ be the c-transform of ψ, that is the function defined by

$$\phi_1(y) := \inf \Big\{ \psi(x) + c(x, y) \mid x \in M \Big\} \qquad \forall y \in M. \tag{3.9}$$

If $y \in \pi^2(S)$, then there is $x \in M$ with $\psi(x) = \phi(y) - c(x, y)$ and $(x, y) \in S \subset \partial_c \psi$, that is

$$\phi(y) = \psi(x) + c(x, y) \leq \psi(z) + c(z, y) \qquad \forall z \in M.$$

Then we get $\phi(y) \leq \phi_1(y)$ for all $y \in \pi^2(S)$. On the other hand, by construction of ψ, we have $\psi(x) \geq \phi(y) - c(x, y)$ for any $x \in M$ and any $y \in \pi^2(S)$. Therefore

$$\phi_1(y) = \phi(y) \qquad \forall y \in \pi^2(S). \tag{3.10}$$

By Proposition 3.5, we have

$$\psi(x) = \sup \Big\{ \phi_1(y) - c(x, y) \mid y \in M \Big\} \qquad \forall x \in M,$$

and by (3.8) and (3.10), we also have

$$\psi(x) = \sup \Big\{ \phi_1(y) - c(x, y) \mid y \in \pi^2(S) \Big\} \qquad \forall x \in M.$$

We claim that ϕ_1 defined by (3.9) satisfies

$$\phi_1(y) = \inf \Big\{ \psi(x) + c(x, y) \mid x \in \pi^1(S) \Big\} \qquad \forall y \in M.$$

If not, there are $\bar{x}, \bar{y} \in M$ and $\delta > 0$ such that

$$\psi(\bar{x}) + c(\bar{x}, \bar{y}) \leq \psi(z) + c(z, \bar{y}) - \delta \qquad \forall z \in \pi^1(S).$$

Taking the infimum in the right-hand side we get

$$\psi(\bar{x}) + c(\bar{x}, \bar{y}) \leq \phi_1(\bar{y}) - \delta.$$

But by construction of ϕ_1, we have $\phi_1(\bar{y}) \leq \psi(\bar{x}) + c(\bar{x}, \bar{y})$. We get a contradiction. It remains to show that both ψ, ψ^c are finite valued and continuous provided c is continuous and S is compact. We claim that under those assumptions, ψ^c is bounded from above on $\pi^2(S)$. Since ψ is not identically $+\infty$, there is $\bar{x} \in M$ with $\psi(\bar{x}) < +\infty$. Since c is continuous and $\pi^2(S)$ is compact, the function $y \mapsto c(\bar{x}, y)$ is bounded on $\pi^2(S)$. Then we deduce that ψ^c is bounded on $\pi^2(y)$. By (3.5), we infer that $\psi(x)$ is finite for any $x \in M$. Let $x \in M$ be fixed and $\{x_k\}_k$ be a sequence converging to x. For every $k > 0$, there is $y_k \in \pi^2(S)$ such that

$$\psi(x_k) \leq \psi^c(y_k) - c(x_k, y_k) + \frac{1}{k}.$$

Then we have for every $k > 0$,

$$\psi(x) \geq \psi^c(y_k) - c(x, y_k) = \psi^c(y_k) - c(x_k, y_k) + c(x_k, y_k) - c(x, y_k)$$

$$\geq \psi(x_k) - \frac{1}{k} + c(x_k, y_k) - c(x, y_k). \qquad (3.11)$$

For every $k > 0$, there is $z_k \in \pi^2(S)$ such that

$$\psi(x) \leq \psi^c(z_k) - c(x, z_k) + \frac{1}{k}.$$

Then we also have for every $k > 0$,

$$\psi(x_k) \geq \psi^c(z_k) - c(x_k, z_k) = \psi^c(z_k) - c(x, z_k) + c(x, z_k) - c(x_k, z_k)$$

$$\geq \psi(x) - \frac{1}{k} + c(x, z_k) - c(x_k, z_k). \qquad (3.12)$$

Let $\mathcal{V}$ be a compact neighborhood of x. The function c is continuous on the compact set $\mathcal{V} \times \pi^2(S)$, hence it is uniformly continuous. We conclude easily from (3.11)–(3.12) that $\psi(x_k)$ tends to $\psi(x)$ as k tends to $+\infty$. In the same way, we can show that ψ is bounded on $\pi^1(S)$ and ψ^c if always valued in $\mathbb{R}$ and continuous. The fact that the infimum and supremum in (3.5)–(3.6) are attained is straigthforward from the continuity of ψ, ψ^c and the compactness of S. $\square$

Corollary 3.8 *Let μ, v be two probability measures on M. Assume that c is continuous and that $Supp(\mu)$ and $Supp(v)$ are compact. Then there is a c-cyclically monotone compact set $\mathscr{S} \subset Supp(\mu) \times Supp(v)$ such that for every $\alpha \in \Pi(\mu, v)$ the following properties are equivalent:*

(i) α is optimal.
(ii) Supp(α) $\subset \mathscr{S}$.

Proof By Theorem 3.3, there is a c-cyclically monotone compact set $\mathscr{S} \subset$ Supp(μ) $\times$ Supp(ν) such that the support of any optimal transport in $\Pi(\mu, \nu)$ is contained in $\mathscr{S}$. Let us show that $\mathscr{S}$ satisfies the equivalence given in the statement of the theorem. First, by construction we have (i) $\Rightarrow$ (ii). By Theorem 3.7, there is a c-pair of potentials with $\mathscr{S} \subset \partial_c \psi$. Then we have

$$\psi^c(y) - \psi(x) = c(x, y) \qquad \forall(x, y) \in \mathscr{S}. \tag{3.13}$$

Furthermore we have

$$\psi^c(y) - \psi(x) \leq c(x, y) \qquad c(x, y) \qquad \forall x, y \in M. \tag{3.14}$$

Let us show that (ii) $\Rightarrow$ (i). Let $\alpha \in \Pi(\mu, \nu)$ be such that Supp(α) $\subset \mathscr{S}$. On the one hand, by (3.13), we have

$$\int_M \psi^c(y)\, d\nu(y) - \int_M \psi(x)\, d\mu(x) = \int_{M \times M} \left(\psi^c(y) - \psi(x) \right) d\alpha(x, y)$$

$$= \int_{M \times M} c(x, y)\, d\alpha(x, y) = C(\alpha).$$

On the other hand, (3.14) yields for every $\alpha' \in \Pi(\mu, \nu)$,

$$\int_M \psi^c(y)\, d\nu(y) - \int_M \psi(x)\, d\mu(x) = \int_{M \times M} \left(\psi^c(y) - \psi(x) \right) d\alpha'(x, y)$$

$$\leq \int_{M \times M} c(x, y)\, d\alpha(x, y) = C(\alpha').$$

This shows that α is optimal. $\square$

Remark 3.7 Let μ, ν be two compactly supported probability measures on M and $c : M \times M \to [0, +\infty)$ be a continuous cost. Actually, the proof of Corollary 3.1 shows that if (ψ, ψ^c) is a c-pair of potentials and α is a transport plan between μ and ν with Supp(α) $\subset \partial_c \psi$, then α is optimal, that is $C_{\mathscr{K}}(\mu, \nu) = C(\alpha)$.

3.3 A Generalized Brenier-McCann Theorem

Throughout this section, we fix a cost $c : M \times M \to [0, +\infty)$ which is assumed to be continuous. Given two compactly supported probability measures μ, ν on M, we know by Theorems 3.3 and 3.7 that there is a c-cyclically monotone compact set $\mathscr{S} \subset$ Supp(μ) $\times$ Supp(ν) which contains the supports of all optimal plans between μ and ν and a c-pair of real-valued continuous potentials (ψ, ψ^c) satisfying

$$\psi(x) = \max\left\{\psi^c(y) - c(x,y) \mid y \in \pi^2(\mathscr{S})\right\} \qquad \forall x \in M, \qquad (3.15)$$

$$\psi^c(y) = \min\left\{\psi(x) + c(x,y) \mid x \in \pi^1(\mathscr{S})\right\} \qquad \forall y \in M, \qquad (3.16)$$

and

$$\mathscr{S} \subset \partial_c\psi. \qquad (3.17)$$

To prove the existence and uniqueness of an optimal transport map, we will show that $\mathscr{S}$ is concentrated on a graph. More precisely, we will prove that for every x outside a μ-negligible set $N \subset M$, the set $\partial_c\psi(x)$ is a singleton.

Theorem 3.9 *Let μ, ν be two probability measures on M. Assume that c is continuous and that $Supp(\mu)$ and $Supp(\nu)$ are compact. Let $\mathscr{S}$ and (ψ, ψ^c) given by Theorems 3.3 and 3.7 as above. Moreover assume that for μ-a.e. $x \in M$, the set $\partial_c\psi(x)$ is a singleton. Then there is a unique optimal transport map from μ to ν. It satisfies*

$$\partial_c\psi(x) = \{T(x)\} \qquad \mu - a.e. \ x \in M. \qquad (3.18)$$

Proof By Theorem 3.1, there is an optimal transport plan α between μ and ν. By assumption, there is a Borel set N such that $\mu(N) = 0$ and for every $x \notin N$, $\partial_c\psi(x)$ is a singleton $\{y_x\}$. Then for every $(x,y) \in Supp(\alpha) \setminus (N \times M)$, we have $(x,y) \in \partial_c\psi$, that is $y = y_x$. Setting $T(x) := y_x$ for μ-a.e. $x \in M$, we get (3.18) and in turn the uniqueness. $\square$

Remark 3.8 We maybe need to make clear what me mean by uniqueness of an optimal transport map. We say that there is a unique optimal transport map from μ to ν if there is uniqueness up to a set of μ-measure zero. That is if T_1 and T_2 are two optimal transport maps from μ to ν, there is a set N with $\mu(N) = 0$ such that $T_1(x) = T_2(x)$ for every $x \notin N$.

We now introduce an assumption on the cost c. For this we need to define the notion of sub-differential. Given an open set $\Omega \subset M$ and a function $f : \Omega \to \mathbb{R}$, we say that $p \in T_x^*M$ is a *sub-differential* for f at $x \in \Omega$ if there is a function $\varphi : \Omega \to \mathbb{R}$ which is differentiable at x with $D_x\varphi = p$ such that (see Fig. 3.3)

$$f(x) = \varphi(x) \quad \text{and} \quad f(y) \geq \varphi(y) \qquad \forall y \in \Omega.$$

We denote by $D_x^- f$ the set of sub-differentials of f at x. In the same way, we say that $p \in T_x^*M$ is a *super-differential* for f at $x \in \Omega$ if there is a function $\varphi : \Omega \to \mathbb{R}$ which is differentiable at x with $D_x\varphi = p$ such that

$$f(x) = \varphi(x) \quad \text{and} \quad f(y) \leq \varphi(y) \qquad \forall y \in \Omega.$$

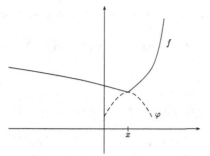

Fig. 3.3 The function φ is a support function from *below* for f at x

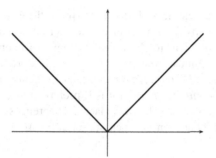

Fig. 3.4 The function $x \mapsto |x|$

We denote by $D_x^+ f$ the set of super-differentials of f at x.

Remark 3.9 If $f : \Omega \to \mathbb{R}$ is differentiable at $x \in \Omega$, then $D_x^- f = D_x^+ f = \{D_x f\}$.

Remark 3.10 The sub-differential and/or the super-differential may not be a singleton. It could be empty or contain several sub-differentials.

For example, the sub-differential of the function $x \mapsto |x|$ at the origin is the interval $[-1, 1]$ while its super-differential is empty (Fig. 3.4).

By (3.15), for every $(x, y) \in \partial_c \psi$ there is a link between the super-differentials of ψ at x and the sub-differentials of the cost c at (x, y). This lead us to the following definition which will be satisfied by variational costs.

Definition 3.10 We say that the cost c satisfies the *sub-TWIST condition* if

$$D_x^- c(\cdot, y_1) \cap D_x^- c(\cdot, y_2) = \emptyset \quad \forall y_1 \neq y_2 \in M, \forall x \in M,$$

where $D_x^-(\cdot, y_i)$ denotes the sub-differential of the function $x \mapsto c(x, y_i)$ at x.

The following result makes the sub-TWIST condition relevant.

Lemma 3.11 *Assume that the cost c satisfies the sub-TWIST condition. Let (ψ, ψ^c) be a c-pair of potentials and $x \in M$ be such that ψ has a non-empty super-differential at x. Then $\partial_c \psi(x)$ is a singleton.*

Proof Argue by contradiction and assume that $y_1 \neq y_2$ both belong to $\partial_c \psi(x)$. Then we have

$$\psi^c(y_i) = \psi(x) + c(x, y_i) \leq \psi(z) + c(z, y_i) \qquad \forall z \in M.$$

Thus, for every $i = 1, 2$,

$$c(z, y_i) \geq -\psi(z) + \psi(x) + c(x, y_i),$$

with equality at $z = x$. Since ψ is super-differentiable at x, we infer that both functions $z \mapsto c(z, y_1)$ and $z \mapsto c(z, y_2)$ share a common sub-differentiable at x. This contradicts the sub-TWIST condition. $\qquad\Box$

By Theorem 3.9 and Lemma 3.11, in order to prove the existence and uniqueness of optimal transport maps from a compactly supported probability measure μ to another one ν, it is sufficient to show that the super-differential of the potential ψ is non-empty for μ-almost every point in M. Such a property can be obtained thanks to Rademacher's Theorem. We recall that a function defined on a smooth manifold is called *Lipschitz in charts* if it is Lipschitz in a set of local coordinates in a neighborhood of any point. The Rademacher Theorem asserts that any function which is Lipschitz in charts on an open subset Ω of M is differentiable almost everywhere in Ω.

Theorem 3.12 *Let $c : M \times M \to [0, +\infty)$ be a cost which is Lipschitz in charts and satisfies the sub-TWIST condition. Let μ, ν be two probability measures with compact support on M. Assume that μ is absolutely continuous with respect to the Lebesgue measure. Then there is existence and uniqueness of an optimal transport map from μ to ν. In fact, there is a c-convex function $\psi : M \to \mathbb{R}$ which is Lipschitz in charts such that*

$$\partial_c \psi(x) = \{T(x)\} \qquad \mu - a.e. \; x \in M. \tag{3.19}$$

Proof By Theorems 3.3 and 3.7 there is a c-cyclically monotone compact set $\mathscr{S} \subset$ $\mathrm{Supp}(\mu) \times \mathrm{Supp}(\nu)$ which contains the supports of all optimal plans between μ and ν together with a c-pair of real-valued continuous potentials (ψ, ψ^c) such that (3.15)–(3.17) are satisfied. In a neighborhood of each $x \in M$, the function ψ is the maximum of a family of functions $x \in \pi^2(S) \mapsto \psi^c(y) - c(x, y)$ with $y \in \pi^2(\mathscr{S})$ which are uniformly Lipschitz (in charts) in the x variable . Therefore, ψ is Lipschitz in charts on M. Since μ is assumed to be absolutely continuous with respect to the Lebesgue measure, Rademacher's Theorem implies that ψ is differentiable and a fortiori super-differentiable μ-a.e. We conclude easily by Theorem 3.9 and Lemma 3.11. $\qquad\Box$

Example 3.7 (Brenier's Theorem) Let $M = \mathbb{R}^n$ and $c : \mathbb{R}^n \times \mathbb{R}^n \to [0, +\infty)$ be the quadratic Euclidean cost or Brenier cost defined by $c(x, y) = |y - x|^2/2$ for any $x, y \in \mathbb{R}^n$. Remembering Example 3.6, we know that c-convex functions are the functions $\psi : \mathbb{R}^n \to \mathbb{R} \cup \{+\infty\}$ such that the function $\psi + |\cdot|^2/2$ is convex. Furthermore, c satisfies the sub-TWIST condition. As a matter of fact, it is smooth and its partial derivative with respect to the x variable is given by

$$\frac{\partial c}{\partial x}(x, y) = x - y \qquad \forall x, y \in \mathbb{R}^n.$$

Therefore $y_1 \neq y_2 \Rightarrow D_x c(\cdot, y_1) \neq D_x c(\cdot, y_2)$. By Theorem 3.12, given a pair of compactly supported probability measures μ, ν in $\mathbb{R}^n$ with μ absolutely continuous with respect to the Lebesgue measure, there is a unique optimal transport map $T : M \to M$ from μ to ν satisfying (3.19) where $\psi : \mathbb{R}^n \to \mathbb{R}$ is a locally Lipschitz c-convex function. Note that for every $x \in \mathbb{R}^n$ where ψ is differentiable at x, we have

$$y \in \partial_c \psi(x) \implies \psi(x) + c(x, y) \leq \psi(z) + c(z, y) \quad \forall z \in \mathbb{R}^n,$$

which means that the derivative of the function $z \mapsto \psi(z) + c(z, y)$ vanishes at $z = x$, that is $y = x + \nabla_x \psi$. Setting $\varphi(x) := \psi(x) + |x|^2/2$ for every $x \in M$, we obtain a convex function such that

$$T(x) = \nabla_x \varphi \qquad \mu - \text{a.e } x \in \mathbb{R}^n.$$

In other terms, the unique optimal transport map from μ to ν is given by the gradient of a convex function.

Example 3.8 Let $M = \mathbb{R}^n$, note that the Monge cost $c : \mathbb{R}^n \times \mathbb{R}^n \to [0, +\infty)$ given by $c(x, y) = |y - x|$ (cf. Examples 3.2, 3.5) is Lipschitz but does not satisfy the sub-TWIST condition. As a matter of fact, we have

$$\frac{\partial c}{\partial x}(x, y) = \frac{x - y}{|x - y|} \qquad \forall x \neq y \in \mathbb{R}^n.$$

This means that $D_x c(\cdot, y_1) = D_x c(\cdot, y_2)$ for any y_1, y_2 such that $y_1 - x$ and $y_2 - x$ are positively colinear. Hence Theorem 3.15 do not apply. In fact, we already saw through Example 3.2 that uniqueness of optimal transport maps does not hold in this context.

Example 3.9 (McCann's Theorem) Let (M, g) be a complete Riemannian manifold. The geodesic distance d_g is Lipschitz in charts on $M \times M$. Define the quadratic geodesic cost or McCann's cost $c : M \times M \to [0, +\infty)$ by

$$c(x, y) := \frac{1}{2}d_g^2(x, y) \qquad \forall x, y \in M.$$

Then c is Lipschitz in charts on $M \times M$ and satisfies the sub-TWIST condition. As a matter of fact, given $x \in M$ and $p \in T_x^* M$ in $D_x^- c(\cdot, y)$ for some $y \in M$, there is a function $\varphi : M \to \mathbb{R}$ which is differentiable at x with $D_x \varphi = p$ such that

$$\frac{1}{2}d_g^2(x, y) = \varphi(x) \quad \text{and} \quad \frac{1}{2}d_g^2(z, y) \geq \varphi(z) \qquad \forall z \in M.$$

Then we argue as in the proof of Lemma 2.15. If we denote by $\bar{\gamma} : [0, 1] \to M$ a minimizing geodesic from y to x, then we obtain that for every curve $\gamma : [0, 1] \to M$ with $\gamma(0) = y$,

$$\frac{1}{2}\text{energy}^g(\gamma) - \varphi(\gamma(1)) \geq 0,$$

with equality for $\gamma = \bar{\gamma}$. As in Lemma 2.15, we infer that there is a unique minimizing geodesic between x and y and that

$$y = \exp_x(-D_x\varphi) = \exp_x(-p),$$

where $\exp_x : T_x^*M \to M$ stands for the exponential map which was defined in Sect. 2.3 (if we use the Riemannian exponential map, we have $y = \exp_x(-\nabla_x^g\varphi)$). The point y is uniquely determined by p, then c satisfies the sub-TWIST condition. Moreover we note that if a potential $\psi : M \to \mathbb{R}$ is (super-) differentiable at $x \in M$ and $y \in \partial_c\psi(x)$, then

$$c(z, y) \geq -\psi(z) + \psi(x) + c(x, y) \qquad \forall z \in M,$$

with equality at $z = x$. Then arguing as above, we deduce that for every pair of compactly supported probability measures μ, ν on M with μ absolutely continuous with respect to the Lebesgue measure, there is a unique optimal transport map T from μ to ν satisfying (3.26) where $\psi : M \to \mathbb{R}$ is a c-convex function which is Lipschitz in charts. By the above discussion, we have

$$T(x) = \exp_x(D_x\psi) \qquad \mu - \text{a.e } x \in M \tag{3.20}$$

and for μ-a.e. $x \in M$ there is a unique minimizing geodesic from x to $T(x)$.

Let M be a smooth connected manifold equipped with a complete sub-Riemannian structure (Δ, g) and whose sub-Riemannian distance is denoted by d_{SR}. In the Sect. 3.4 our purpose is now to study the Monge problem for the sub-Riemannian quadratic cost, that is for the cost $c : M \times M \to [0, +\infty)$ defined by

$$c(x, y) := \frac{1}{2}d_{SR}(x, y)^2 \qquad \forall x, y \in M.$$

As we saw before, in order to obtain existence and uniqueness results for optimal transport maps, it is convenient to be able to show that super-differentials of potentials are non-empty almost everywhere and that some sub-TWIST condition is satisfied by the cost function. The sub-TWIST condition follows immediately from Lemma 2.15. So we just have to deal with regularity issues of c-convex functions. In the case of compactly supported probability measures, regularity properties of Kantorovitch potentials can be obtained from the regularity of the cost. We develop this approach in the Sect. 3.4 by showing that under additional assumptions the sub-Riemannian distance is Lipschitz and even locally semiconcave outside the diagonal.

Remark 3.11 As explained above, if M equipped with a SR structure for which the cost $c = d_{SR}^2$ is Lipschitz on $M \times M$, then for every pair of compactly supported probability measures μ, ν on M with μ absolutely continuous with respect to the Lebesgue measure, there is a unique optimal transport map T from μ to ν which can be expressed as

$$T(x) = \exp_x(D_x \psi) \qquad \mu - \text{a.e } x \in M, \tag{3.21}$$

where $\psi : M \to \mathbb{R}$ is a c-convex function which is Lipschitz in charts.

3.4 Optimal Transport on Ideal and Lipschitz SR Structures

Ideal SR structures. Let (Δ, g) be a sub-Riemannian structure of rank $m \leq n$ on M. We call it *ideal* if it is complete and has no non-trivial minimizing singular curves. We recall that this implies that for every $x \neq y \in M$, any minimizing geodesic $\gamma : [0, 1] \to M$ joining x to y is regular. By the results of the Chap. 2, all minimizing geodesics are smooth and projections of normal extremals of the Hamiltonian geodesic equation. We recall that D denotes the diagonal of $M \times M$, that is, the set of all pairs of the form (x, x) with $x \in M$. Sub-Riemannian distances of ideal SR structures are locally semiconcave outside the diagonal.

A function $f : \Omega \to \mathbb{R}$, defined on the open set $\Omega \subset M$, is called *locally semiconcave* on Ω if for every $x \in \Omega$ there exist a neighborhood Ω_x of x and a smooth diffeomorphism $\varphi_x : \Omega_x \to \varphi_x(\Omega_x) \subset \mathbb{R}^n$ such that $f \circ \varphi_x^{-1}$ is locally semiconcave on the open subset $\tilde{\Omega}_x = \varphi_x(\Omega_x) \subset \mathbb{R}^n$. By the way, we recall that the function $\tilde{f} : \tilde{\Omega} \to \mathbb{R}$, defined on the open set $\tilde{\Omega} \subset \mathbb{R}^n$, is locally semiconcave on $\tilde{\Omega}$ if for every $\bar{x} \in \tilde{\Omega}$ there exist $C, \delta > 0$ such that

$$\mu f(y) + (1 - \mu) f(x) - f(\mu x + (1 - \mu) y)$$
$$\leq \mu (1 - \mu) C |x - y|^2 \quad \forall \mu \in [0, 1], \ \forall x, y \in B(\bar{x}, \delta).$$

This is equivalent to say that the function $\tilde{f}$ can be written locally as

$$\tilde{f}(x) = \left(\tilde{f}(x) - C|x|^2 \right) + C|x|^2 \quad \forall x \in B(\bar{x}, \delta),$$

with $\tilde{f}(x) - C|x|^2$ concave, that is as the sum of a concave function and a smooth function. Note that every locally semiconcave function is locally Lipschitz on its domain, and thus, by Rademacher's Theorem, it is differentiable almost everywhere on its domain.

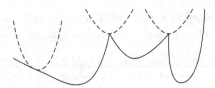

Fig. 3.5 Graph of a semiconcave function

The following result is useful to prove the local semiconcavity of a given function (Fig. 3.5).

Lemma 3.13 *Let $f : \Omega \to \mathbb{R}$ be a function defined on an open set $\Omega \subset \mathbb{R}^n$. Assume that for every $\bar{x} \in \Omega$ there exist a neighborhood $\mathscr{V} \subset \Omega$ of $\bar{x}$ and a positive real number σ such that, for every $x \in \mathscr{V}$, there is $p_x \in \mathbb{R}^n$ such that*

$$u(y) \leq u(x) + \langle p_x, y - x \rangle + \sigma |y - x|^2 \quad \forall y \in \mathscr{V}.$$

Then the function u is locally semiconcave on Ω.

Proof (Proof of Lemma 3.13) Let $\bar{x} \in \Omega$ be fixed and $\mathscr{V}$ be the neighborhood given by assumption. Without loss of generality, we can assume that $\mathscr{V}$ is an open ball $\mathscr{B}$. Let $x, y \in \mathscr{B}$ and $\mu \in [0, 1]$. The point $\hat{x} := \mu x + (1 - \mu)y$ belongs to $\mathscr{B}$. By assumption, there exists $\hat{p} \in \mathbb{R}^n$ such that

$$u(z) \leq u(\hat{x}) + \langle \hat{p}, z - \hat{x} \rangle + \sigma |z - \hat{x}|^2 \quad \forall z \in \mathscr{B}.$$

Hence we easily get

$$\mu u(y) + (1 - \mu)u(x) \leq u(\hat{x}) + \mu\sigma |x - \hat{x}|^2 + (1 - \mu)\sigma |y - \hat{x}|^2$$
$$\leq u(\hat{x}) + \left(\mu(1 - \mu)^2\sigma + (1 - \mu)\mu^2\sigma \right) |x - y|^2$$
$$\leq u(\hat{x}) + 2\mu(1 - \mu)\sigma |x - y|^2,$$

and the conclusion follows.   □

Remark 3.12 Thanks to Lemma 3.13, a way to prove that a given function $f : \Omega \to \mathbb{R}$ is locally semiconcave on Ω is to show that for every $x \in \Omega$ we can put a C^2 support function φ on the graph of u at x with a uniform control of the C^2 norm of φ.

Outside the diagonal, sub-Riemannian distances of ideal SR structures enjoy the same kind of regularity as Riemannian distances.

Theorem 3.14 *Let (Δ, g) be an ideal sub-Riemannian structure on M. Then the SR distance is continuous on $M \times M$ and locally semiconcave on $M \times M \setminus D$. In particular, d_{SR} is Lipschitz in charts on $M \times M \setminus D$.*

Proof The continuity of d_{SR} follows from Proposition 1.13. To prove the local semi-concavity, we proceed as explained in Remark 3.12. Let us fix $(x, y) \in M \times M \setminus D$ and $\gamma \in \Omega_{\Delta}^{x,1}$ be a minimizing geodesic joining x to y. There is an open neighborhood $\mathcal{V}$ of $\gamma([0, 1])$ in M and an orthonormal family $\mathcal{F}$ (with respect to the metric g) of m smooth vector fields $X^1, \ldots, X^m$ such that

$$\Delta(z) = \text{Span}\left\{X^1(z), \ldots, X^m(z)\right\} \qquad \forall z \in \mathcal{V}.$$

Taking a change of coordinates if necessary, we may assume that $\mathcal{V}$ is an open subset of $\mathbb{R}^n$. Furthermore, there is a control $u^\gamma \in L^2([0, 1]; \mathbb{R}^m)$ such that

$$\dot{\gamma}(t) = \sum_{i=1}^{m} u_i^\gamma(t) X^i(\gamma(t)) dt \qquad \text{a.e. } t \in [0, 1].$$

Since u^γ is regular, there are $v^1, \ldots v^n$ in $L^2([0, 1]; \mathbb{R}^m)$ such that the linear operator

$$\mathbb{R}^n \longrightarrow \mathbb{R}^n$$

$$\alpha \longmapsto \sum_{i=1}^{m} \alpha_i D_{u^\gamma} E_{\mathcal{F}}^{x,1}\left(v^i\right)$$

is invertible. Define locally $\mathscr{F} : \mathbb{R}^n \to \mathbb{R}^n$ by

$$\mathscr{F} : \mathbb{R}^n \times \mathbb{R}^n \longrightarrow \mathbb{R}^n \times \mathbb{R}^n$$

$$(z, \alpha) \longmapsto \left(z, E_{\mathcal{F}}^{z,1}\left(u^\gamma + \sum_{i=1}^{m} \alpha_i v^i\right)\right).$$

This mapping is well-defined and C^2 in a neighborhood of $(x, 0)$. Moreover it satisfies

$$\mathscr{F}(x, 0) = (x, y),$$

and its differential at $(x, 0)$ is invertible. Hence by the Inverse Function Theorem, there are an open ball $\mathscr{B}$ centered at (x, y) in $\mathbb{R}^n \times \mathbb{R}^n$ and a function $\mathscr{G} : \mathscr{B} \to \mathbb{R}^n \times \mathbb{R}^n$ of class C^2 such that

$$\mathscr{F} \circ \mathscr{G}(z, w) = (z, w) \qquad \forall (z, w) \in \mathscr{B}.$$

Denote by α^{-1} the second component of $\mathscr{G}$. From the definition of the sub-Riemannian energy between two points, we infer that for any $(z, w) \in \mathscr{B}$ we have

$$e_{SR}(z, w) \leq \left\| u^\gamma + \sum_{i=1}^{m} \left(\alpha^{-1}(z, w)\right)_i v_i \right\|_{L^2}^2.$$

Set

$$\phi^{x,y}(z,w) := \left\| u^\gamma + \sum_{i=1}^{m} \left(\alpha^{-1}(z,w) \right)_i \right\|_{L^2} \quad \forall(z,w) \in \mathscr{B}.$$

We conclude that, there is a function $\phi^{x,y}$ of class C^2 such that $d_{SR}(z,w) \le \phi^{x,y}(z,w)$ for any (z,w) in a neighborhood of (x,y) in $M \times M$, and $d_{SR}(x,y) = \phi^{x,y}(x,y)$. By compactness, the C^2 norms of the functions $\phi^{x,y}$ are uniformly bounded. As a matter of fact, from Remark 2.2 we know that the set of minimizing geodesics from x to y is compact with respect to the uniform topology; any sequence of minimizing geodesics $\{\gamma_k\}_k$ from x_k to y_k converges uniformly to a minimizing geodesic from x to y. We also know (see Remark 2.1) that if we cover the set of minimizing curves from x to y by a finite number of open tubes admitting orthonormal frames, then minimizing control converge in L^2. We conclude easily. $\quad\square$

Remark 3.13 The above arguments can be used to prove the following result. Let (Δ, g) be a sub-Riemannian structure of rank $m < n$ on M. Assume that it is complete and that there is an open set $\Omega \subset M \times M$ such that for every $(x,y) \in \Omega$ with $x \ne y$, any minimizing geodesic between x and y is regular. Then d_{SR} is locally semiconcave on $\Omega \setminus D$.

Remark 3.14 Any SR structure of rank $m = n$, that is any Riemannian structure on M is ideal, see Remarks 1.9, 2.3.

Lipschitz SR structures. Let (Δ, g) be a sub-Riemannian structure of rank $m < n$ on M. We call it *Lipschitz* if it is complete and if the sub-Riemannian distance function is Lipschitz in charts on $M \times M$ outside the diagonal (or equivalently if the sub-Riemannian energy is Lipschitz in charts on $M \times M \setminus D$). A particular case of Lipschitz SR structures is given by ideal SR structures. The aim of the present section is to provide a weaker sufficient condition for a complete SR structure to be Lipschitz. According to Theorem 2.22, a horizontal path $\gamma : [0,1] \to M$ will be called a *Goh path* if it admits an abnormal lift $\psi : [0,1] \to \Delta^\perp$ which annihilates $[\Delta, \Delta]$, that is, an abnormal lift $\psi = (\gamma, p) : [0,1] \to T^*M$ (in local coordinates, see Proposition 1.11 and the subsequent remarks) such that for every local parametrization of Δ by smooth vector fields $X^1, \ldots, X^m$ in a neighborhood of $\gamma([0,1])$, we have

$$p(t) \cdot [X^i, X^j](\gamma(t)) = 0 \quad \forall t \in [0,1], \, \forall i,j = 1, \ldots, m.$$

Of course, the above definition does not depend upon the parametrization.

Theorem 3.15 *Let (Δ, g) be a complete sub-Riemannian structure on M, assume that any sub-Riemannian minimizing geodesic joining two distinct points in M is not a Goh path. Then, the SR structure (Δ, g) is Lipschitz.*

Proof Let us fix $(x,y) \in M \times M \setminus D$ and $\gamma \in \Omega_\Delta^{x,1}$ a minimizing geodesic joining x to y. As before, denote by $\mathscr{F} = \{X^1, \ldots, X^m\}$ an orthonormal family of vector fieldsd along $\gamma([0, 1])$ and by u^γ the control associated with γ. Two cases may appear:

First case: $\bar{u} := u^\gamma$ is not singular.
Then by the arguments given in the proofs of Lemma 2.18 and Theorem 3.14, there are $\delta, K > 0$ such that

$$e_{SR}(x, z) \leq e_{SR}(x, y) + K|z - y| \quad \forall z \in B(y, \delta).$$

Since any control which is close enough to $\bar{u}$ is regular, there is $\bar{\varepsilon} > 0$ such that for every $u \in L^2([0, 1]; \mathbb{R}^m)$ satisfying

$$\|u - \bar{u}\|_{L^2} < \bar{\varepsilon}, \quad e_{SR}\left(x, E_{\mathscr{F}}^{x,1}(u)\right) = \|u\|_{L^2},$$

there holds

$$e_{SR}(x, z) \leq e_{SR}\left(x, E_{\mathscr{F}}^{x,1}(u)\right) + 2K \left|z - E_{\mathscr{F}}^{x,1}(u)\right|, \tag{3.22}$$

for every $z \in B\left(E_{\mathscr{F}}^{x,1}(u), \delta/2\right)$.

Second case: $\bar{u} = u^\gamma$ is singular.
By Theorem 2.20, we have necessarily

$$\mathrm{ind}_-\left(\lambda^*\left(D_{\bar{u}}^2 E_{\mathscr{F}}^{x,1}\right)_{|\mathrm{Ker}(D_{\bar{u}} E_{\mathscr{F}}^{x,1})}\right) = +\infty, \tag{3.23}$$

for all $\lambda \in \mathrm{Im}\left(D_{\bar{u}} E_{\mathscr{F}}^{x,1}\right)^\perp \setminus \{0\}$. Recall that $C : L^2([0, 1]; \mathbb{R}^m)$ is defined by

$$C(u) := \|u\|_{L^2}^2 \quad \forall u \in L^2([0, 1]; \mathbb{R}^m).$$

Let $E_0 \subset L^2([0, 1]; \mathbb{R}^m)$ be a vector space such that

$$E_0 + \mathrm{Ker}\left(D_{\bar{u}} E_{\mathscr{F}}^{x,1}\right) = L^2([0, 1]; \mathbb{R}^m).$$

Set
$$E := E_0 \oplus \left(\mathrm{Ker}\left(D_{\bar{u}} E_{\mathscr{F}}^{x,1}\right) \cap \mathrm{Ker}\left(D_{\bar{u}} C\right)\right) \quad \text{and} \quad F := \left(E_{\mathscr{F}}^{x,1}\right)_{|\{\bar{u}\}+E}.$$

By construction, $D_{\bar{u}} E_{\mathscr{F}}^{x,1}$ and $D_{\bar{u}} F$ have the same image in $\mathbb{R}^n$ and E_0 has finite dimension. Then by (3.23), we have

$$\mathrm{ind}_-\left(\lambda^*\left(D_{\bar{u}}^2 F\right)_{|\mathrm{Ker}(D_{\bar{u}} F)}\right) = +\infty,$$

for all $\lambda \in \mathrm{Im}\,(D_{\bar{u}}F)^{\perp} \setminus \{0\}$. We can apply Theorem B.4 to the function F. Hence there are $c > 0$, $\bar{\varepsilon} \in (0, 1)$ such that for every $\varepsilon \in (0, \bar{\varepsilon})$ and every $z \in B\left(F(\bar{u}), c\varepsilon^2\right)$, there are $w_1, w_2 \in L^2([0, 1]; \mathbb{R}^m)$ such that

$$z = F(\bar{u} + w_1 + w_2) \tag{3.24}$$

and

$$w_1 \in \mathrm{Ker}(D_{\bar{u}}F), \quad \|w_1\|_{L^2} < \varepsilon, \quad \|w_2\|_{L^2} < \varepsilon^2. \tag{3.25}$$

Let $z \in B(y, c\varepsilon^2)$ with $|z - y| = c\varepsilon^2/2$. Then there are $w_1, w_2 \in L^2([0, 1]; \mathbb{R}^m)$ such that (3.24)–(3.25) are satisfied. Set $u := \bar{u} + w_1 + w_2$. Then we have

$$z = E_{\mathscr{F}}^{x,1}(u),$$

and (note that $\mathrm{Ker}(D_{\bar{u}}F) \subset \mathrm{Ker}(D_{\bar{u}}C)$),

$$\begin{aligned}
e_{SR}(x, z) \leq C(u) &\leq C(\bar{u}) + D_{\bar{u}}C \cdot (w_1 + w_2) + \|w_1 + w_2\|_{L^2}^2 \\
&= e_{SR}(x, y) + D_{\bar{u}}C \cdot w_2 + \|w_1 + w_2\|_{L^2}^2 \\
&\leq e_{SR}(x, y) + 2\|\bar{u}\|_{L^2}\varepsilon^2 + (\varepsilon + \varepsilon^2)^2 \\
&\leq e_{SR}(x, y) + \left(\frac{4\|\bar{u}\|_{L^2} + 8}{c}\right)|z - y|.
\end{aligned}$$

Proceeding as in the proof of Theorem B.4, we can show that the above estimate holds in a neighborhood of $\bar{u}$, that is (taking $c > 0$, $\bar{\varepsilon} \in (0, 1)$ smaller if necessary) for every $\varepsilon \in (0, \bar{\varepsilon})$, for every $u \in L^2([0, 1]; \mathbb{R}^m)$, and every $z \in \mathbb{R}^n$ with

$$\|u - \bar{u}\|_{L^2} < \varepsilon, \quad \left|z - E_{\mathscr{F}}^{x,1}(u)\right| < c\varepsilon^2,$$

there are $w_1, w_2 \in L^2([0, 1]; \mathbb{R}^m)$ such that

$$z = E_{\mathscr{F}}^{x,1}(u + w_1 + w_2)$$

and

$$w_1 \in \mathrm{Ker}(D_u C), \quad \|w_1\|_{L^2} < \varepsilon, \quad \|w_2\|_{L^2} < \varepsilon^2.$$

This shows that for every $u \in L^2([0, 1]; \mathbb{R}^m)$ satisfying

$$\|u - \bar{u}\|_{L^2} < \bar{\varepsilon}, \quad e_{SR}\left(x, E_{\mathscr{F}}^{x,1}(u)\right) = \|u\|_{L^2},$$

there holds

$$e_{SR}(x, z) \le e_{SR}\left(x, E_{\mathscr{F}}^{x,1}(u)\right) + \left(\frac{4\|u\|_{L^2} + 8}{c}\right)\left|z - E_{\mathscr{F}}^{x,1}(u)\right|, \qquad (3.26)$$

for every $z \in B\left(E_{\mathscr{F}}^{x,1}(u), c\bar{\varepsilon}/4\right)$.

Let us explain how to conclude by compactness. Let $x \in M$ and $\mathscr{B}$ a compact set in M such that $\{x\} \times \mathscr{B} \cap D = \emptyset$ be fixed. Denote by $\mathscr{S}$ the set of all $y \in \mathscr{B}$ such that there is at least one singular minimizing geodesic between x and y. The set $\mathscr{S}$ is a compact subset of $\mathscr{B}$, and the set of singular minimizing geodesic between x and a point in $\mathscr{S}$ is compact with respect to the uniform topology. Then by the previous observation (second case) together with a compactness argument (see Remarks 2.1, 2.2), we infer that an inequality of the form (3.26) holds for any minimizing control u which is close enough to a control corresponding to a singular minimizing geodesic joining x to a point in $\mathscr{S}$. Denote by $\mathscr{S}'$ the set of y in $\mathscr{B}$ corresponding to such controls. By construction, any minimizing geodesic from x to a point in $\mathscr{B} \setminus \mathscr{S}'$ is regular. Actually, it is far from being singular. Then by the arguments given in the first case together with compactness arguments, an inequality of the form (3.22) holds for any $y (= E_{\mathscr{F}}^{x,1}(u))$ in $\overline{\mathscr{B} \setminus \mathscr{S}'}$. In that way, we prove that $e_{SR}(x, \cdot)$ (or equivalently $d_{SR}(x, \cdot)$) is locally Lipschitz in $M \setminus \{x\}$. The same proof shows that e_{SR} is indeed uniformly locally Lipschitz with respect to one variable. We conclude easily. $\qquad \square$

Remark 3.15 The above arguments can be used to prove the following result. Let (Δ, g) be a sub-Riemannian structure of rank $m < n$ on M. Assume that it is complete and that there is an open set $\Omega \subset M \times M$ such that for every $(x, y) \in \Omega$ with $x \ne y$, no minimizing geodesic between x and y is a Goh path. Then d_{SR} is Lipschitz in charts on $\Omega \setminus D$.

Remark 3.16 Note that if the path γ is constant on $[0, 1]$, it is a Goh path if and only if there is a differential form $p \in T_{\gamma(0)}^* M$ satisfying

$$p \cdot X^i(\gamma(0)) = p \cdot [X^i, X^j](\gamma(0)) = 0 \qquad \forall i, j = 1, \dots, m,$$

where $X^1, \dots, X^m$ is as above a parametrization of Δ in a neighborhood of $\gamma(0)$. The above proof shows that if Δ is 2-generating then e_{SR} is Lipschitz in charts on $M \times M$.

Remark 3.17 If a SR structure (Δ, g) on M is Lipschitz, then for every $x \in M$, the exponential mapping $\exp_x$ is onto. In fact, for every y there is a minimizing geodesic joining x to y which is normal. This can be shown by the arguments which were given at the end of the proof of Theorem 2.14.

A Brenier-McCann Theorem on Lipschitz SR structures. Before stating our existence and uniqueness result for Lipschitz SR structures, we introduce a definition.

Definition 3.16 Given a c-convex function $\psi : M \to \mathbb{R}$, we call *moving set* $\mathscr{M}^\psi$ and *static set* $\mathscr{S}^\psi$ respectively the sets defined as follows:

$$\mathscr{M}^\psi := \left\{ x \in M \mid x \notin \partial_c \psi(x) \right\},$$

$$\mathscr{S}^\psi := M \setminus \mathscr{M}^\phi = \left\{ x \in M \mid x \in \partial_c \psi(x) \right\}.$$

As shown by the following result, under classical assumptions on the measures and Lipschitzness of the sub-Riemannian structure, static points do not move while moving points obey a transportation law of the form (3.20)–(3.21).

Theorem 3.17 *Let (Δ, g) be a Lipschitz sub-Riemannian structure on M and μ, ν be two compactly supported probability measures on M. Assume that μ is absolutely continuous with respect to the Lebesgue measure. Then there is existence and uniqueness of an optimal transport map from μ to ν for the SR quadratic cost $c : M \times M \to [0, +\infty)$ defined by*

$$c(x, y) := \frac{1}{2} d_{SR}^2(x, y) \quad \forall x, y \in M.$$

In fact, there is a continuous c-convex function $\psi : M \to \mathbb{R}$ such that the following holds:

(i) *$\mathscr{M}^\psi$ is open, and ψ is Lipschitz in charts on $\mathscr{M}^\psi$. In particular ψ is differentiable μ-a.e. in $\mathscr{M}^\psi$.*
(ii) *For μ-a.e, $x \in \mathscr{S}^\psi$, $\partial_c \psi(x) = \{x\}$.*

In particular, there exists a unique optimal transport map defined μ-a.e. by

$$T(x) := \begin{cases} \exp_x (D_x \psi) & \text{if } x \in \mathscr{M}^\psi, \\ x & \text{if } x \in \mathscr{S}^\psi, \end{cases}$$

and for μ-a.e. $x \in M$ there exists a unique minimizing geodesic between x and $T(x)$.

Proof Let $\mathscr{S} \subset \mathrm{Supp}(\mu) \times \mathrm{Supp}(\nu)$ and (ψ, ψ^c) be respectively the c-cyclically monotone set and the c-pair of potentials satisfying (3.15)–(3.17). Since the sets $\mathrm{Supp}(\mu), \mathrm{Supp}(\nu)$ are assumed to be compact, both ψ, ψ^c are indeed continuous and the supremum and infimum in (3.15)–(3.16) are attained. We check easily that $x \in M$ belongs to $\mathscr{S}^\psi$ if and only if $\psi(x) = \psi^c(x)$. Then $\mathscr{M}^\psi$ coincides with the set

$$\left\{ x \in M \mid \psi(x) \neq \psi^c(x) \right\} = \left\{ x \in M \mid \psi(x) > \psi^c(x) \right\},$$

which is open by continuity of ψ and ψ^c. Let us now prove that ψ is Lipschitz in charts in an open neighborhood of $\mathscr{M}^\psi \cap \mathrm{Supp}(\mu)$. Let $x \in \mathscr{M}^\psi$ be fixed. Since $x \notin \partial_c \psi(x)$ and $\psi_c(x)$ is closed in M (by continuity of ψ, ψ_c and compactness of

$\mathscr{S}$), there is $r > 0$ such that $d_{SR}(x, y) > 2r$ for any $y \in \partial_c \psi(x)$. In addition, since the set $\partial_c \psi$ is closed in $M \times M$ (again by continuity of ψ, ψ_c and compactness of $\mathscr{S}$), there exists a neighborhood $\mathscr{V}_x$ of x which is included in $\mathscr{M}^{\psi}$ such that

$$d_{SR}(z, w) \geq r \qquad \forall z \in \mathscr{V}_x, \quad \forall w \in \partial_c \psi(z).$$

Let $\psi_{x,r} : M \to \mathbb{R}$ be the function defined by

$$\psi_{x,r}(z) := \sup\left\{\psi^c(y) - \frac{1}{2}d_{SR}^2(z, y) \mid y \in \pi^2(\mathscr{S}), \, d_{SR}(z, y) \geq r\right\}.$$

By construction, ψ coincides with $\psi_{x,r}$ on $\mathscr{V}_x$. By assumption, d_{SR} is Lipschitz in charts outside the diagonal, then by compactness of $\mathscr{S}$ we deduce that $\psi_{x,r}$ is Lipschitz in charts. In conclusion Ψ is Lipschitz in charts on $\mathscr{M}^{\psi}$ and (i) is proved.

To prove (ii), we observe that it suffices to prove the result for x belonging to an open set $\mathscr{V} \subset M$ on which the horizontal distribution $\Delta(x)$ is parametrized by a orthonormal family a smooth vector fields $\mathscr{F} = \{X^1, \ldots, X^m\}$. In fact, up to working in charts, we can assume that $\mathscr{V}$ is a convex subset of $\mathbb{R}^n$ where the C^2-norms of the X^i's are bounded. Let us fix a compact ball $\mathscr{B}$ in $\mathscr{V}$ and show that (ii) holds for μ-a.e. $x \in \mathscr{B}$.

Recall that the Hamiltonian $H : \mathscr{V} \times (\mathbb{R}^n)^* \to \mathbb{R}$ which is associated to our sub-Riemannian structure is defined by (see Chap. 2)

$$H(x, p) := \frac{1}{2}\sum_{i=1}^{m}\left(p \cdot X^i(x)\right)^2 \qquad \forall (x, p) \in \mathscr{V} \times (\mathbb{R}^n)^*.$$

For every $p \in (\mathbb{R}^n)^* \setminus \{0\}$, denote by Π_p the linear hyperplane in $\mathbb{R}^n$ which is orthogonal to p, that is

$$\Pi_p := \left\{v \in \mathbb{R}^n \mid p \cdot v = 0\right\}.$$

From Lemma 2.11 and its proof, for every $\bar{x} \in \mathscr{V}$ and every $\bar{p} \in (\mathbb{R}^n)^*$ with $H(\bar{x}, \bar{p}) \neq 0$, there is $\rho > 0$ such that the Dirichlet problem

$$\begin{cases} H(x, D_x S(x)) = H(\bar{x}, \bar{p}), \\ S_{|\bar{x}+\Pi_{\bar{p}}} = 0, \end{cases} \tag{3.27}$$

admits a solution of class C^1 on the ball $B(\bar{x}, \rho)$. We leave the reader to check that the radius ρ depends "continuously" on $\bar{x}$, $H(\bar{x}, p)$ and $|p|$ ($|p|$ denotes the Euclidean norm of p). Then, by compactness of $\mathscr{B}$ there is a function

$$\rho : (0, +\infty) \times (0, +\infty) \longrightarrow (0, \infty)$$

which is decreasing in the first variable and increasing in the second variable such that for every $\bar{x} \in \mathscr{B}$ and every $\bar{p} \in (\mathbb{R}^n)^*$ with $H(\bar{x}, \bar{p}) \neq 0$, the solution to (3.27) is

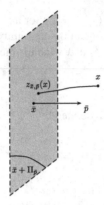

Fig. 3.6 Characteristics of $S_{\bar{x},\bar{p}}$

defined on the open ball $B\left(\bar{x}, \rho(H(\bar{x}, \bar{p}), |\bar{p}|)\right)$. For any $\bar{x}, \bar{p}$ satisfying the previous assumptions, we denote by

$$S_{\bar{x},\bar{p}} \,:\, B\left(\bar{x}, \rho\big(H(\bar{x},\bar{p}), |\bar{p}|\big)\right) \longrightarrow \mathbb{R}$$

the solution to the Dirichlet problem (3.27), with $\rho_{\bar{x},\bar{p}} := \rho(H(\bar{x}, \bar{p}), |\bar{p}|)$. The functions $S_{\bar{x},\bar{p}}$ being constructed by the method of characteristics (see Proof of Lemma 2.11), the following result holds (note that the parametrization of characteristics that we use in the statement of Lemma 3.18 differs from the one which is used to construct $S_{\bar{x},\bar{p}}$, see last statement).

Lemma 3.18 *There is a function*

$$\tau \,:\, (0, +\infty) \times (0, +\infty) \longrightarrow (0, +\infty)$$

which is increasing in the first variable and decreasing in the second variable such that the following property holds (see Fig. 3.6):
For every $\bar{x} \in \mathcal{B}$, for every $\bar{p} \in (\mathbb{R}^n)^*$ with $H(\bar{x}, \bar{p}) \neq 0$, and every $x \in B\left(\bar{x}, \rho_{\bar{x},\bar{p}}/2\right)$ there are

$$z_{\bar{x},\bar{p}}(x) \in \left(\bar{x} + \Pi_{\bar{p}}\right) \cap B\left(\bar{x}, \rho_{\bar{x},\bar{p}}\right)$$

and $t_{\bar{x},\bar{p}}(x) \in \left(-\tau\big(H(\bar{x},\bar{p}), |\bar{p}|\big), \tau\big(H(\bar{x},\bar{p}), |\bar{p}|\big)\right)$

such that
$$x = \gamma_{\bar{x},\bar{p}}\big(t_{\bar{x},\bar{p}}(x); z_{\bar{x},\bar{p}}(x)\big)$$

where (we set $\tau_{\bar{x},\,\bar{p}} := \tau\big(H(\bar{x},\,\bar{p}), |\bar{p}|\big)$)

$$\left(\gamma_{\bar{x},\bar{p}}\left(\cdot\,; z_{\bar{x},\bar{p}}(x)\right), p_{\bar{x},\bar{p}}\left(\cdot\,; z_{\bar{x},\bar{p}}(x)\right)\right) \; : \; \left(-\tau_{\bar{x},\bar{p}}, \tau_{\bar{x},\bar{p}}\right) \longrightarrow \mathcal{V} \times (\mathbb{R}^n)^*$$

is the solution to the Hamiltonian system

$$
\begin{cases}
\dot{\gamma}_{\bar{x},\bar{p}}\left(t; z_{\bar{x},\bar{p}}(x)\right) = \dfrac{\partial H}{\partial p}\left(\gamma_{\bar{x},\bar{p}}\left(t; z_{\bar{x},\bar{p}}(x)\right), p_{\bar{x},\bar{p}}\left(t; z_{\bar{x},\bar{p}}(x)\right)\right) \\[2mm]
\dot{p}_{\bar{x},\bar{p}}\left(t; z_{\bar{x},\bar{p}}(x)\right) = -\dfrac{\partial H}{\partial x}\left(\gamma_{\bar{x},\bar{p}}\left(t; z_{\bar{x},\bar{p}}(x)\right), p_{\bar{x},\bar{p}}\left(t; z_{\bar{x},\bar{p}}(x)\right)\right),
\end{cases}
$$

with

$$\gamma_{\bar{x},\bar{p}}\left(0; z_{\bar{x},\bar{p}}(x)\right) = z_{\bar{x},\bar{p}}(x) \quad and \quad p_{\bar{x},\bar{p}}\left(0; z_{\bar{x},\bar{p}}(x)\right) = \bar{p}.$$

In particular, $\gamma_{\bar{x},\bar{p}}$ is an horizontal path joining $z_{\bar{x},\bar{p}}(x)$ to x which satisfies

$$H\left(\gamma_{\bar{x},\bar{p}}\left(t; z_{\bar{x},\bar{p}}(x)\right), p_{\bar{x},\bar{p}}\left(t; z_{\bar{x},\bar{p}}(x)\right)\right) = H\left(z_{\bar{x},\bar{p}}(x), \bar{p}\right) \qquad \forall t \in \left(-\tau_{\bar{x},\bar{p}}, \tau_{\bar{x},\bar{p}}\right).$$

For every $x \in \mathcal{V}$, we denote by $\Delta^\perp(x)$ the set of $p \in (\mathbb{R}^n)^*$ such that $H(x, p) \neq 0$. Pick a sequence $\{(x_k, p_k)\}_k$ of $\mathcal{B} \times (\mathbb{R}^n)^*$ which is a dense subset of

$$\left\{(x, p) \in \mathcal{B} \times (\mathbb{R}^n)^* \mid p \in \Delta^\perp(x)\right\}.$$

and set for every k,

$$\rho_k := \rho_{x_k,p_k}, \qquad \tau_k := \tau_{x_k,p_k}, \qquad t_k(\cdot) := t_{x_k,p_k}(\cdot),$$

$$z_k(\cdot) := z_{x_k,p_k}(\cdot), \qquad \gamma_k(\cdot,\cdot) := \gamma_{x_k,p_k}(\cdot,\cdot), \qquad p_k(\cdot,\cdot) := p_{x_k,p_k}(\cdot,\cdot).$$

The following result is a consequence of the Lipschitz regularity of the sub-Riemannian distance along horizontal paths together with Rademacher's theorem.

Lemma 3.19 *There is a set N of Lebesgue measure zero in $\mathcal{V}$ such that for every $x \in \mathcal{B} \setminus N$ and any k, the following property holds:*

$$x \in B(x_k, \rho_k/2) \; and \; x = \gamma_k\left(t; z_k(x)\right)$$
$$\Longrightarrow s \mapsto \psi\left(\gamma_k\left(s; z_k(x)\right)\right) \; is \; differentiable \; at \; t.$$

Proof (Proof of Lemma 3.19) Let k be fixed. By construction, all the curves $\gamma_k(\cdot\,; z)$ (with $z \in \left(\bar{x} + \Pi_{p_k}\right) \cap B(x_k, \rho_k)$) are horizontal with respect to the distribution (we may assume without loss of generality that the curves $\gamma_k(\cdot\,; z)$ are defined on $(-\tau_k, \tau_k)$ for all $z \in \left(\bar{x} + \Pi_{p_k}\right) \cap B(x_k, \rho_k)$). The potential ψ is expressed as

$$\psi(x) = \max\left\{\psi^c(y) - \frac{1}{2}d_{SR}^2(x, y) \mid y \in \pi^2(\mathscr{S})\right\} \qquad \forall x \in M,$$

with ψ^c continuous and $\pi^2(\mathcal{S})$ compact. Hence, given $\bar{s} \in (-\tau_k, \tau_k)$, there is $\bar{y} \in \pi^2(\mathcal{S})$ such that

$$\psi\left(\gamma_k(\bar{s}; z)\right) = \psi^c(\bar{y}) - \frac{1}{2}d_{SR}^2\left(\gamma_k(\bar{s}; z), \bar{y}\right).$$

Then we have for every $s \in (-\tau_k, \tau_k)$,

$$\begin{aligned}
\psi\left(\gamma_k(s; z)\right) &\geq \psi^c(\bar{y}) - \frac{1}{2}d_{SR}^2\left(\gamma_k(s; z), \bar{y}\right) \\
&\geq \psi^c(\bar{y}) - d_{SR}^2\left(\gamma_k(s; z), \gamma_{k,l}(\bar{s}; z)\right) - d_{SR}^2\left(\gamma_k(\bar{s}; z), \bar{y}\right) \\
&\geq \psi\left(\gamma_k(\bar{s}; z)\right) - 2H(z, p_k)\left|s - \bar{s}\right|^2 \\
&\geq \psi\left(\gamma_k(\bar{s}; z)\right) - 4\tau_k H(z, p_k)\left|s - \bar{s}\right|.
\end{aligned}$$

This shows that each function $s \mapsto \psi\left(\gamma_k(s; z)\right)$ is locally Lipschitz on its domain. By Rademacher's theorem, we infer that it is differentiable almost everywhere on $(-\tau_k, \tau_k)$. Since the paths $\gamma_k(\cdot; z)$ with $z \in \left(x_k + \Pi_{p_k}\right) \cap B\left(x_k, \rho_k\right)$ laminate a set which is bigger than the ball $B(x_k, \rho_k/2)$ in a continuous way, Fubini's theorem implies the existence of a negligeable set $N_{k,l}$ such that the property stated in the lemma holds for k. We conclude by setting $N = \cup N_k$.   □

Before starting the proof of (ii), we need a last result giving an estimates on the deviation of normal geodesics. For every (x, p), we denote by $\left(\gamma_{x,p}, p_{x,p}\right) := \left(\gamma_{x,p}(\cdot; x), p_{x,p}(\cdot; x)\right)$, the solution of the Hamiltonian system starting at (x, p); it is defined on the interval $(-\tau(H(x, p), |p|), \tau(H(x, p), |p|))$.

Lemma 3.20 *There is a function*

$$C : (0, +\infty) \times (0, +\infty) \longrightarrow (0, +\infty)$$

which is decreasing in the first variable and increasing in the second variable such that the following property holds:
For every $h, R > 0$, every k, and every $(x, p) \in \mathcal{B} \times (\mathbb{R}^n)^*$ satisfying

$$H\left(x_k, p_k\right), H(x, p) > h, \quad \left|p_k\right|, |p| < R, \quad x \in B\left(x_k, \rho_k/2\right), \tag{3.28}$$

one has

$$\left|\gamma_k\left(t_k(x) + s; z_k(x)\right) - \gamma_{x,p}(s)\right| \leq C(h, R)\left|p_k\left(t_k(x); z_k(x)\right) - p\right| s, \tag{3.29}$$

for every $s \in (-\tau(h, R), \tau(h, R)) \cap (-t_k(x) - \tau(h, R), -t_k(x) + \tau(h, R))$.

Proof (Proof of Lemma 3.20) Since the C^1-norms of the X^i's are bounded on $\mathcal{V}$, there is an increasing function $P : (0, +\infty) \to (0, +\infty)$ such that the solutions to our Hamiltonian system starting from a pair (x, p) with $x \in \mathcal{B}, H(x, p) > h$ and

$|p| < R$ remains in the set $\mathcal{V} \times B(0, P(R))$ on the interval $(-\tau(h, R), \tau(h, R))$ (note that since H is constant along the Hamiltonian trajectories, the solutions remains in the set $\{H(x, p) > h\}$). Now, considering Lipschitz constants of the Hamiltonian vector field on the "cylinder" $\mathcal{V} \times B(0, P(R))$ (the C^2-norms of the X^i's are bounded on $\mathcal{V}$) and using Gronwall's Lemma (see Appendix A), we prove easily the existence of an increasing function $C : [0, +\infty) \to [0, +\infty)$ such that

$$\left| \gamma_k\left(t_k(x) + s; z_k(x)\right) - \gamma_{x,p}(s) \right| + \left| p_k\left(t_k(x) + s; z_k(x)\right) - p_{x,p}(s) \right|$$
$$\leq C(R) \left| p_k\left(t_k(x); z_k(x)\right) - p \right|, \tag{3.30}$$

for every $h, R > 0$, every k, and every $(x, p) \in \mathcal{B} \times (\mathbb{R}^n)^*$ satisfying (3.28), and every $s \in (-\tau(h, R), \tau(h, R)) \cap (-t_k(x) - \tau(h, R), -t_k(x) + \tau(h, R))$. Let us denote by I the latter interval and set

$$u(s) := \left| \gamma_k\left(t_k(x) + s; z_k(x)\right) - \gamma_{x,p}(s) \right| \qquad \forall s \in I.$$

Considering again the Lipschitz constants of the Hamiltonian vector field that we always denote by K, we obtain formally for every s,

$$u(s) = \left| \int_0^s \frac{\partial H}{\partial p} \left(\gamma_k\left(t_k(x) + r; z_k(x)\right), p_k\left(t_k(x) + r; z_k(x)\right) \right) - \frac{\partial H}{\partial p} \left(\gamma_{x,p}(r), p_{x,p}(r) \right) dr \right|$$
$$\leq K \int_0^s u(r) \, dr + K \int_0^s \left| p_k\left(t_k(x) + r; z_k(x) - p_{x,p}(r) \right| \, dr,$$

which by (3.30) gives

$$u(s) \leq K \int_0^s u(r) \, dr + K \int_0^s C(R) \left| p_k\left(t_k(x); z_k(x)\right) - p \right| \, dr.$$

Gronwall's Lemma (see Lemma A.1) concludes the proof. $\qquad\square$

We are now ready to prove that for every $x \in \mathcal{B} \setminus N$, we have $\partial_c \psi(x) = \{x\}$. Fix $x \in \mathcal{B} \setminus N$ and argue by contradiction, that is assume that there is $\bar{y} \neq x$ such that $\bar{y} \in \partial^c \psi(x) \setminus \{x\}$. Then we have (remembering Remark 3.6)

$$\psi(x) + c(x, \bar{y}) \leq \psi(z) + c(z, \bar{y}) \qquad \forall z \in M,$$

which can be written as

$$\psi(x) - \psi(z) \leq \frac{1}{2} d_{SR}^2(z, \bar{y}) - \frac{1}{2} d_{SR}^2(x, \bar{y}) \qquad \forall z \in M. \tag{3.31}$$

Since d_{SR} is Lipschitz in charts outside the diagonal, there is a normal minimizing geodesic joining x to $\bar{y}$ (see Remark 3.17), that is there is $p \in T_x^* M$ such that $\exp_x(p) = \bar{y}$ and $d_{SR}(x, y)^2 = 2H(x, p) \neq 0$. Note that since x belongs to $\partial_c \psi(x)$, we have

$$\psi(x) = \psi(x) + c(x, x) \le \psi(z) + c(z, x) \qquad \forall z \in \mathcal{V}.$$

Set $h := H(x, p)/2$, $R := 2|p|$ and pick k such that

$$H(x_k, p_k) > h, \quad |p_k|, |p| < R, \quad x \in B(x_k, \rho_k/2).$$

Applying the previous inequality with $z = \gamma_k(t_k(x) + s; z_k(x))$ and s small yields

$$
\begin{aligned}
\psi\left(\gamma_k\left(t_k(x); z_k(x)\right)\right) &= \psi(x) \\
&\le \psi\left(\gamma_k\left(t_k(x) + s; z_k(x)\right)\right) + \frac{1}{2}d_{SR}^2\left(\gamma_k\left(t_k(x) + s; z_k(x)\right), x\right) \\
&\le \psi\left(\gamma_k\left(t_k(x) + s; z_k(x)\right)\right) + H\left(z_k(x), p_k\right) s^2,
\end{aligned}
$$

because $\gamma_k(\cdot; z_k(x))$ is an horizontal path joining $x = \gamma_k(t_k(x); z_k(x))$ to the point $\gamma_k(t_k(x) + s; z_k(x))$ of length $s\sqrt{2H(z_k(x), p_k)}$. Since x does not belong to N, the function

$$s \longmapsto \psi\left(\gamma_k\left(t_k(x) + s; z_k(x)\right)\right)$$

is differentiable at $s = 0$. Then Lemma 3.19 together with the previous inequality allows us to write

$$\frac{d}{ds}\psi\left(\gamma_k\left(t_k(x) + s; z_k(x)\right)\right)\Big|_{s=0} = 0. \qquad (3.32)$$

Since d_{SR} is Lipschitz outside the diagonal and $\bar{y} \ne x$, there are $\rho, K > 0$ such that

$$\left|d_{SR}^2(z, \bar{y}) - d_{SR}^2(z', \bar{y})\right| \le K|z' - z| \qquad \forall z, z' \in B(x, \rho).$$

Then applying (3.31) with $z = \gamma_k\left(t_k(x) + s; z_k(x)\right)$ and s small and using (3.29) yields (as in Lemma 3.20, $\gamma_{x,p}$ denotes the geodesic starting at x with initial covector p, note that $\gamma_{x,p}(s)$ belongs to $\mathcal{V}$ for small s)

$$
\begin{aligned}
\psi(x) &- \psi\left(\gamma_k\left(t_k(x) + s; z_k(x)\right)\right) \\
&\le \frac{1}{2}d_{SR}^2\left(\gamma_k\left(t_k(x) + s; z_k(x)\right), \bar{y}\right) - \frac{1}{2}d_{SR}^2(x, \bar{y}) \\
&\le \frac{K}{2}\left|\gamma_k\left(t_k(x) + s; z_k(x)\right) - \gamma_{x,p}(s)\right| + \frac{1}{2}d_{SR}^2\left(\gamma_{x,p}(s), \bar{y}\right) - \frac{1}{2}d_{SR}^2(x, \bar{y}) \\
&\le \frac{KC(h, R)}{2}\left|p_k\left(t_k(x); z_k(x)\right) - p\right|s + \frac{1}{2}(1 - s)^2 d_{SR}^2(x, \bar{y}) - \frac{1}{2}d_{SR}^2(x, \bar{y}) \\
&= \left(\frac{KC(h, R)}{2}\left|p_k\left(t_k(x); z_k(x)\right) - p\right| - d_{SR}^2(x, \bar{y})\right)s + \frac{d_{SR}^2(x, y)}{2}s^2.
\end{aligned}
$$

The quantity

$$\frac{KC(h, R)}{2} \left| p_k\big(t_k(x); z_k(x)\big) - p \right|$$

tends to 0 as (x_k, p_k) tends to (x, p). We infer that for (x_k, p_k) close enough to (x, p), the derivative of the function $s \mapsto \psi\big(\gamma_k(t_k(x) + s; z_k(x))\big)$ cannot be zero. This contradicts (3.32).

It remains to prove the formula for $T(x)$ and the uniqueness of minimizing geodesic between x and $T(x)$ μ-almost everywhere. We need to show that

$$\partial^c \psi(x) \cap \mathrm{Supp}(v) = \exp_x\left(\frac{1}{2} D_x \psi\right)$$

for all $x \in \mathcal{M}^\psi \cap \mathrm{Supp}(\mu)$ where ψ is differentiable, which is the case for μ-almost every $x \in \mathcal{M}^\psi$ by assertion (i) and Rademacher's theorem. This is a consequence of Lemma 2.15 applied to the function $z \mapsto -\psi(z) + \psi^c(y)$ at the point x with $y \in \partial \psi_c(x)$. Moreover, again by Lemma 2.15, the geodesic from x to $T(x)$ is unique for μ-a.e. $x \in \mathcal{M}^\psi \cap \mathrm{Supp}(\mu)$. Since $T(x) = x$ for $x \in \mathcal{S}^\psi \cap \mathrm{Supp}(\mu)$, the geodesic is clearly unique also in this case. □

Remark 3.18 If the sub-Riemannian structure is assumed to be ideal, then the potential ψ can be shown to be locally semiconcave on the moving set.

Remark 3.19 The above arguments show that Theorem 3.17 remains true under more general assumptions. Let (Δ, g) be a complete sub-Riemannian structure on M and μ, v be two compactly supported probability measures in M with μ absolutely continuous with respect to the Lebesgue measure. Assume that there are two open sets $\Omega_1, \Omega_2 \subset M$ with

$$\mu(M \setminus \Omega_1) = 0 \quad \text{and} \quad \mathrm{Supp}(v) \subset \Omega_2$$

such that the sub-Riemannian distance is Lipschitz in charts on $(\Omega_1 \times \Omega_2) \setminus D$. Then there is existence and uniqueness of an optimal transport map with respect to the sub-Riemannian quadratic cost.

3.5 Back to Examples

We conclude the present chapter with a list of examples for which we have existence and uniqueness of optimal transport maps for the SR quadratic cost, that is the cost $c : M \times M \to [0, +\infty)$ defined by

$$c(x, y) := \frac{1}{2} d_{SR}^2(x, y) \quad \forall x, y \in M.$$

Given a cost function, we shall say that the Monge problem is *well-posed*, if we have existence and uniqueness of optimal transport maps from an absolutely continuous compactly supported measure to a compactly supported measure. All the examples that we review below have already been encoutenred within the text.

Fat distributions. Recall (see Example 1.15) that a distribution Δ on M is called *fat* if, for every $x \in M$ and every section X of Δ with $X(x) \neq 0$, there holds

$$T_x M = \Delta(x) + [X, \Delta](x),$$

where

$$[X, \Delta](x) := \left\{ [X, Z](x) \mid Z \text{ section of } \Delta \right\}.$$

We saw that fat distributions do not admit non-trivial singular horizontal paths. This means that any complete sub-Riemannian structure associated with a fat distribution is ideal. In conclusion, by Theorem 3.17, the Monge problem for any sub-Riemannian structure associated with a fat distributions is well-posed.

Two-generating distributions. A distribution Δ is called *two-generating* if

$$T_x M = \Delta(x) + [\Delta, \Delta](x) \qquad \forall x \in M.$$

Two-generating distributions do not admit Goh paths (see Example 2.1). By Theorem 3.17, the Monge problem for any sub-Riemannian structure associated with a two-generating distributions is well-posed.

Totally nonholonomic distributions on three-dimensional manifolds. Assume that M has dimension 3, that Δ is a nonholonomic rank-two distribution on M, and define

$$\Sigma_\Delta := \left\{ x \in M \mid \Delta(x) + [\Delta, \Delta](x) \neq \mathbb{R}^3 \right\}.$$

The set Σ_Δ is called the *singular set* or the *Martinet set* of Δ.

Proposition 3.21 *Let Δ be a totally nonholonomic distribution on a three-dimensional manifold. Then, the set Σ_Δ is a closed subset of M which is countably 2-rectifiable. Moreover, a non-trivial horizontal path $\gamma : [0, 1] \to M$ is singular if and only if it is included in Σ_Δ.*

Proof The first part will follow from Proposition 3.22 while the second part has already been proved in Example 1.17.   □

Proposition 3.21 implies that for any pair $(x, y) \in M \times M$ (with $x \neq y$) such that x or y does not belong to Σ_Δ, any sub-Riemannian minimizing geodesic between x and y is nonsingular. Moreover Σ_Δ has Lebesgue measure zero. As a consequence, by Remarks 3.13 and 3.19, the Monge problem is well-posed.

Medium-fat distributions. The distribution Δ is called *medium-fat* if, for every $x \in M$ and every vector field X on M such that $X(x) \in \Delta(x) \setminus \{0\}$, there holds

$$T_x M = \Delta(x) + [\Delta, \Delta](x) + [X, [\Delta, \Delta]](x).$$

As shown in Example 2.1, medium-fat distributions do not admit non-trivial Goh paths. As a consequence, the Monge problem for sub-Riemannian structures involving medium-fat distributions is well-posed.

Codimension-one nonholonomic distributions. Let M have dimension n and Δ be a nonholonomic distribution of rank $n-1$. As in the case of nonholonomic distributions on three-dimensional manifolds, we can define the singular set associated to the distribution as

$$\Sigma_\Delta := \left\{ x \in M \mid \Delta(x) + [\Delta, \Delta](x) \neq T_x M \right\}.$$

The following result holds.

Proposition 3.22 *If Δ is a nonholonomic distribution of rank $n - 1$, then the set Σ_Δ is a closed subset of M which is countably $(n - 1)$-rectifiable. Moreover, any Goh path is contained in Σ_Δ.*

Proof The fact that Σ_Δ is a closed subset of M is obvious. Let us prove that it is countably $(n - 1)$-rectifiable. Since it suffices to prove the result locally, we can assume that we have

$$\Delta(x) = \text{Span}\left\{ X^1(x), \ldots, X^{n-1}(x) \right\} \qquad \forall x \in \mathcal{V},$$

where $\mathcal{V}$ is an open neighborhood of the origin in $\mathbb{R}^n$. Moreover, doing a change of coordinates if necessary, we can also assume that (with coordinates $(x_1, \ldots, x_n)$)

$$X^i = \partial_{x_i} + \alpha_i(x)\, \partial_{x_n} \qquad \forall i = 1, \ldots, n - 1,$$

where each $\alpha_i : \mathcal{V} \longrightarrow \mathbb{R}$ is a C^∞ function satisfying $\alpha_i(0) = 0$. Hence, for any $i, j \in \{1, \ldots n - 1\}$, we have

$$[X^i, X^j] = \left[\left(\frac{\partial \alpha_j}{\partial x_i} - \frac{\partial \alpha_i}{\partial x_j} \right) + \left(\frac{\partial \alpha_j}{\partial x_n} \alpha_i - \frac{\partial \alpha_i}{\partial x_n} \alpha_j \right) \right] \partial_{x_n},$$

and so

$$\Sigma_\Delta = \left\{ x \in \mathscr{V} \mid \left(\frac{\partial \alpha_j}{\partial x_i} - \frac{\partial \alpha_i}{\partial x_j} \right) + \left(\frac{\partial \alpha_j}{\partial x_n} \alpha_i - \frac{\partial \alpha_i}{\partial x_n} \alpha_j \right) = 0 \right.$$

$$\left. \forall i, j \in \{1, \ldots, n-1\} \right\}.$$

For every tuple $I = (i_1, \ldots, i_k) \in \{1, \ldots, n-1\}^k$ we denote by X^I the smooth vector field constructed by Lie brackets of $X^1, X^2, \ldots, X^{n-1}$ as follows,

$$X^I = \left[X^{i_1}, \left[X^{i_2}, \ldots, \left[X^{i_{k-1}}, X^{i_k} \right] \ldots \right] \right].$$

We call $k = \mathrm{length}(I)$ the length of the Lie bracket X^I. Since Δ is totally nonholonomic, there is some positive integer r such that

$$\mathbb{R}^n = \mathrm{Span}\left\{ X^I(x) \mid \mathrm{length}(I) \leq r \right\} \quad \forall x \in \mathscr{V}.$$

It is easy to see that, for every I such that $\mathrm{length}(I) \geq 2$, there is a smooth function $g_I : \mathscr{V} \to \mathbb{R}$ such that

$$X^I(x) = g_I(x) \partial_{x_n} \quad \forall x \in \mathscr{V}.$$

Defining the sets A_k as

$$A_k := \left\{ x \in \mathscr{V} \mid g_I(x) = 0 \ \ \forall I \text{ such that } \mathrm{length}(I) \leq k \right\},$$

we have

$$\Sigma_\Delta = \bigcup_{k=2}^{r} (A_k \setminus A_{k+1}).$$

By the Implicit Function Theorem, it is easy to see that each set $A^k \setminus A^{k+1}$ can be covered by a countable union of smooth hypersurfaces. Indeed assume that some given x belongs to $A_k \setminus A_{k+1}$. This implies that there is some $J = (j_1, \ldots, j_{k+1})$ of length $k+1$ such that $g_J(x) \neq 0$. Set $I = (j_2, \ldots, j_{k+1})$. Since $g_I(x) = 0$, we have

$$g_J(x) = \left(\frac{\partial g_I}{\partial x_{j_1}}(x) + \frac{\partial g_I}{\partial x_n}(x) \alpha_{j_1}(x) \right) \neq 0.$$

Hence, either $\frac{\partial g_I}{\partial x_{j_1}}(x) \neq 0$ or $\frac{\partial g_I}{\partial x_n}(x) \neq 0$.

Consequently, we deduce that we have the following inclusion

$$A^k \setminus A^{k+1} \subset \bigcup_{\mathrm{length}(I)=k} \left\{ x \in \mathscr{V} \mid \exists i \in \{1, \ldots, n\} \text{ such that } \frac{\partial g_I}{\partial x_i}(x) \neq 0 \right\}.$$

We conclude easily.

The fact that any Goh path is contained in Σ_Δ is obvious.   □

As a consequence by Remarks 3.13 and 3.19, the Monge problem for sub-Riemannian structures involving codimension one distributions is well-posed.

Rank-two distributions in dimension four. Let (M, Δ, g) be a complete sub-Riemannian manifold of dimension four, and let Δ be a *regular* rank-two distribution, that is satisfying

$$T_x M$$
$$= \operatorname{Span}\left\{X^1(x), X^2(x), [X^1, X^2](x), \left[X^1, [X^1, X^2]\right](x), \left[X^2, [X^1, X^2]\right](x)\right\}$$

for any local parametrization $\mathscr{F} = \{X^1, X^2\}$ of the distribution. In Example 1.19, we saw that there is a smooth horizontal vector field X on M such that the singular horizontal paths γ parametrized by arc-length are exactly the integral curves of X, i.e. the curves satisfying

$$\dot\gamma(t) = X(\gamma(t)).$$

For every $x \in M$, denote by $\mathscr{O}(x)$ the orbit of x by the flow of X and set

$$\Omega := \left\{(x, y) \in M \times M \mid y \notin \mathscr{O}(x)\right\}.$$

According to Remark 3.13, the following result holds:

Proposition 3.23 *Under the assumption above, the function d_{SR} is locally semiconcave in the interior of Ω.*

The above result allow us to obtain existence and uniqueness of optimal transport maps in certain cases. Let us consider the distribution given in Example 1.18, that is the distribution Δ in $\mathbb{R}^4$ spanned by the vector fields

$$X^1 = \partial x_1, \qquad X^2 = \partial x_2 + x_1 \partial x_3 + x_3 \partial x_4.$$

As shown in Example 1.18, an horizontal path $\gamma : [0, 1] \to \mathbb{R}^4$ is singular if and only if it satisfies, up to reparameterization by arc-length,

$$\dot\gamma(t) = X^1(\gamma(t)) \qquad \forall t \in [0, 1].$$

By the above proposition, we deduce that, for any complete metric g on $\mathbb{R}^4$, the sub-Riemannian distance function d_{SR} locally semiconcave on the set

$$\Omega = \left\{(x, y) \in \mathbb{R}^4 \times \mathbb{R}^4 \mid (y - x) \notin \operatorname{Span}\{e_1\}\right\},$$

where e_1 denotes the first vector in the canonical basis of $\mathbb{R}^4$. Consequently, for any pair of compactly supported probability measures μ, ν on M such that μ is absolutely continuous with respect to the Lebesgue measure and

$$\text{Supp}(\mu \times \nu) \subset \Omega,$$

the Monge problem is well-posed.

3.6 Notes and Comments

In 1781, Monge's original work [18] was concerned with the moving of soil that was modelized as an optimal transport problem consisting in minimizing the transportation cost

$$\int_{\mathbb{R}^3} |T(x) - x| \, d\mu(x), \tag{3.33}$$

between continuous distributions of mass. The Monge problem was rediscovered several decades later, in 1942, by Kantorovitch [15] who proved a duality theorem to study the relaxed form of the problem (which is by now referred as Kantorovitch problem). We refer the reader to the textbook [24] by Villani and references therein for an historical account on the optimal transport theory.

The Kantorovitch duality theorem which is not precisely stated in the present monograph appears through Theorem 3.7 and Corollary 3.1. Actually, our presentation of the theory leading to existence and uniqueness of optimal transport maps closely follows the one of Gangbo and McCann in [13]. For sake of simplicity, we restrict our attention to transportation problems between compactly supported probability measures from a smooth manifold into itself with continuous costs. Most of the results of Sects. 3.1–3.2 remain true in the more general context of lower semi-continuous costs on the product of two Polish spaces and non-compactly supported probability measures. We refer the reader to Villani's monograph [24] for general statements.

As seen through Example 3.1, transport maps may not exist. In fact, Pratelli [20] proved that transport maps do exist as soon as the initial measure is assumed to be non-atomic. The Prokhrorov Theorem which is used in the proof of Theorem 3.1 can be found in Billingsley's book [4]. Theorem 3.7 extends a result by Rockafellar [23] about the sub-differentials of convex functions. The sub-TWIST condition introduced in Sect. 3.3 is a natural extension of the classical TWIST condition (see [24]). Thanks to Lemma 2.15, many costs obtained in a variational way do satisfy the sub-TWIST condition. This is the case of the quadratic Euclidean cost appearing in Example 3.7, or of the quadratic geodesic cost appearing in Example 3.9. In fact, Examples 3.7–3.9 refer respectively to theorems by Brenier [5] and McCann [16]. This type of result can be developed further by considering locally Lipschitz costs associated with problems of calculus of variations involving Tonelli Lagrangians (see [3]) or even with some optimal control problems (see [1]). As seen in Example 3.2, minimizers of the original Monge problem with cost $c(x, y) = |y - x|$ in $\mathbb{R}^n$ may not be unique. However, existence of optimal transport maps can be proved, see [24] and references therein.

The study of Monge-type problems in sub-Riemannian geometry began with a paper by Ambrosio and Rigot [2] about the transportation problem in the Heisenberg group. Then, Agrachev and Lee [1] extended the well-posedness result of Ambrosio-Rigot to the case of sub-Riemannian quadratic costs which are Lipschitz in charts on $M \times M$ (see Remark 3.11). Then, Figalli and the author [12] removed the assumption of Lipschitzness on the diagonal; this is Theorem 3.17. We observe that our proof of assertion (ii) differs from the original proof in [12] which was based on a Pansu-Rademacher Theorem. All these results are concerned with SR quadratic costs (that is $c = d_{SR}^2$). As in the Euclidean case, the Monge problem for the non-quadratic cost $c = d_{SR}$ does not enjoy uniqueness. Using techniques developed by Champion and De Pascale [8], De Pascale and Rigot [10] obtained an existence result for the classical Monge problem in the Heisenberg group.

The local semiconcavity of some SR distances outside the diagonal is demonstrated in Theorem 3.14. Such regularity is fundamental and sometimes necessary. First, it shows that distances of ideal sub-Riemannian structures share the same type of properties as Riemannian distances, at least outside the diagonal. It can be useful to get Sard's theorems and as a consequence regularity properties of sub-Riemannian spheres, see [21]. Then, the semiconcavity of the cost allows to consider probability measures which do not charge rectifiable sets and hence not necessarily absolutely continuous, see [24]. Finally, semiconcavity of the cost may be transfered to potentials (see Remark 3.18) and then permit to get a Monge-Ampère-like equation (see Remark 3.1). This latter consequence is due to a famous theorem by Alexandrov (see [11]) which states that locally semiconvace functions are two times differentiable almost everywhere. We refer the reader to [12] for further details on sub-Riemannian Monge-Ampère equations, to [6, 22] for further details on semiconcave SR distances, and to the Cannarsa-Sinestrari's book [7] for an detailed exposition on semiconcavity.

Our list of examples already appeared in [12] which indeed contained an additional example about generic sub-Riemannian structures. Chitour, Jean and Trélat [9] proved that generic SR structures of rank ≥ 3 do not admit singular curves. By Theorems 3.14 and 3.17, this shows that the Monge problem for generic SR structures of rank ≥ 3 is well-posed. We refer the reader to [12] and references therein for further details.

We do not know if the Monge problem (for the SR quadratic cost) is well-posed for general sub-Riemannian structures. The method presented in this chapter requires regularity properties for d_{SR}. According to the Mitchell ball-box theorem (see [14, 17, 19]), the sub-Riemannian distance is always locally Hölder in charts. In Chap. 2, we saw that given a complete sub-Riemannian structure and $x \in M$ the function

$$y \in M \longmapsto d_{SR}(x, y)$$

is Lipschitz in charts on a dense subset of M. We do not know if this set has necessarily full Lebesgue measure in M (note that the Sard Conjecture that we mentioned in Sect. 2.6 would imply such a result). Anyway, such a result would not be sufficient to prove the well-posedness of Monge problem for general sub-Riemannian structures.

References

1. Agrachev, A., Lee, P.: Optimal transportation under nonholonomic constraints. Trans. Amer. Math. Soc. **361**(11), 6019–6047 (2009)
2. Ambrosio, L., Rigot, S.: Optimal transportation in the Heisenberg group. J. Funct. Anal. **208**(2), 261–301 (2004)
3. Bernard, P., Buffoni, B.: Optimal mass transportation and Mather theory. J. Eur. Math. Soc. **9**(1), 85–121 (2007)
4. Billingsley, P.: Convergence of Probability Measures, 2nd edn. John Wiley & Sons Inc., New York (1999)
5. Brenier, Y.: Polar factorization and monotone rearrangement of vector-valued functions. Comm. Pure Appl. Math. **44**, 375–417 (1991)
6. Cannarsa, P., Rifford, L.: Semiconcavity results for optimal control problems admitting no singular minimizing controls. Ann. Inst. H. Poincaré Anal. Non Linéaire **25**(4), 773–802 (2008)
7. Cannarsa, P., Sinestrari, C.: Semiconcave functions, Hamilton-Jacobi equations, and optimal control. Progress in Nonlinear Differential Equations and their Applications, vol. 58. Birkhäuser, Boston (2004)
8. Champion, T., De Pascale, L.: The Monge problem in $\mathbb{R}^d$. Duke Math. J. **157**(3), 551–572 (2011)
9. Chitour, Y., Jean, F., Trélat, E.: Genericity results for singular curves. J. Differ. Geom. **73**(1), 45–73 (2006)
10. De Pascale, L., Rigot, S.: Monge's transport problem in the Heisenberg group. Adv. Calc. Var. **4**(2), 195–227 (2011)
11. Evans, L.C., Gariepy, R.: Measure Theory and Fine Properties of Functions. CRC Press, Boca Raton, FL (1992)
12. Figalli, A., Rifford, L.: Mass transportation on sub-Riemannian manifolds. Geom. Funct. Anal. **20**(1), 124–159 (2010)
13. Gangbo, W., McCann, R.J.: The geometry of optimal transportation. Acta Math. **177**(2), 113–161 (1996)
14. Jean, F.: Control of nonholonomic systems and sub-Riemannian geometry. Lectures given at the CIMPA School "Géométrie sous-riemannienne", Beirut, Lebanon (2012)
15. Kantorovitch, L.: On the translocation of masses. C.R. (Doklady) Acad. Sci. URSS **37**, 199–201 (1942)
16. McCann, R.J.: Polar factorization of maps on Riemannian manifolds. Geom. Funct. Anal. **11**(3), 589–608 (2001)
17. Mitchell, J.: On Carnot-Carathéodory metrics. J. Differ. Geom. **21**(1), 35–45 (1985)
18. Monge, G.: Mémoire sur la théorie des déblais et des remblais. Histoire de l'Académie Royale des Sciences de Paris, pp. 666–704 (1781)
19. Montgomery, R.: A tour of subriemannian geometries, their geodesics and applications. Mathematical Surveys and Monographs, vol. 91. American Mathematical Society, Providence, RI (2002)
20. Pratelli, A.: On the equality between Monge's infimum and Kantorovitch's minimum in optimal mass transportation. Ann. Inst. H. Poincaré Probab. Statist. **43**(1), 1–13 (2007)
21. Rifford, L.: À propos des sphères sous-riemanniennes. Bull. Belg. Math. Soc. Simon Stevin **13**(3), 521–526 (2006)

22. Rifford, L., Trélat, E.: On the stabilization problem for nonholonomic distributions. J. Eur. Math. Soc. **11**(2), 223–255 (2009)
23. Rockafellar, R.T.: Characterization of the subdifferentials of convex functions. Pacific J. Math. **17**, 497–510 (1966)
24. Villani, C.: Optimal Transport, Old and New. Springer-Verlag, Heidelberg (2008)

Appendix A
Ordinary Differential Equations

We recall here without proofs basic facts on ordinary differential equations. For further details, we refer the reader to the textbook [1].

A function $f : [a, b] \to \mathbb{R}^n$ is said to be *absolutely continuous*, if for each $\varepsilon > 0$, there exists $\delta > 0$ such that for each family of disjoints intervals $\{]a_i, b_i[\}_{i \in \mathbb{N}}$ included in $[a, b]$, and satisfying

$$\sum_{i \in \mathbb{N}} b_i - a_i < \delta,$$

we have

$$\sum_{i \in \mathbb{N}} |f(b_i) - f(a_i)| < \varepsilon.$$

Any absolutely continuous function is continuous. In fact, a function $f : [a, b] \to \mathbb{R}^n$ is absolutely continuous if and only if it is differentiable almost everywhere on $[a, b]$, its derivative $\dot{f}(t) := \frac{d}{dt} f(t)$ is integrable with respect to the Lebesgue measure on $[a, b]$, and we have for each $t \in [a, b]$,

$$\frac{df}{dt} = f(a) + \int_a^t \frac{d}{dt} f(s) ds \quad \forall t \in [a, b].$$

A function $f : [a, b] \to \mathbb{R}^n$ is called *absolutely continuous with square integrable derivative* if it is absolutely continuous on $[a, b]$ and satisfies

$$\dot{f} \in L^2([a, b]; \mathbb{R}^n).$$

Let M be a smooth manifold without boundary of dimension $n \geq 2$. A function $f : [a, b] \to M$ is called absolutely continuous (resp. absolutely continuous with square integrable derivative) if it is absolutely continuous (resp. absolutely continuous with

L. Rifford, *Sub-Riemannian Geometry and Optimal Transport*,
SpringerBriefs in Mathematics, DOI: 10.1007/978-3-319-04804-8,
© The Author(s) 2014

square integrable derivative) in charts. Such a notion does not depend on the atlas chosen to cover M.

The Gronwall lemma is a key tool to obtain estimates involving solutions of differential equations.

Lemma A.1 (Gronwall's Lemma) *Let $\varepsilon > 0$, $\alpha : [0, \varepsilon] :\to \mathbb{R}$ be a continuous function, and $\beta \in L^1([0; \varepsilon], \mathbb{R})$. Assume that $u : [0, \varepsilon] \to \mathbb{R}$ is a continuous function satisfying*

$$u(t) \leq \alpha(t) + \int_0^t \beta(s)u(s)ds \quad \forall t \in [0, \varepsilon].$$

Then there holds

$$u(t) \leq \alpha(t) + e^{\int_0^t \beta(s)ds} \int_0^t e^{-\int_0^s \beta(r)dr} \beta(s)\alpha(s)\, ds \quad \forall t \in [0, \varepsilon].$$

If, in addition, α is nondecreasing, then

$$u(t) \leq \alpha(t)e^{\int_0^t \beta(s)ds} \quad \forall t \in [0, \varepsilon].$$

Let $I \subset \mathbb{R}$ be an open interval, Ω be an open subset of $\mathbb{R}^n$, and $f : I \times \Omega \to \mathbb{R}^n$ be a function satisfying the following property:

(H_{CP}) For every $x \in \Omega$, there exist $\delta > 0$, a locally integrable function $c : I \to [0, +\infty)$, and a nondecreasing function $\omega : [0, +\infty) \to [0, +\infty)$ with $\omega(h) \to 0$ as $h \to 0$ such that

$$|f(t, y) - f(t, z)| \leq c(t)\omega(|y - z|) \quad \text{and} \quad |f(t, y)| \leq c(t)$$

for almost all $t \in I$ and all $y, z \in B(x, \delta)$.

Given $(t_0, x_0) \in I \times \Omega$, our aim is to solve locally the following Cauchy problem

$$\dot{x}(t) = f(t, x(t)), \quad \text{a.e.} t, \quad x(t_0) = x_0. \tag{A.1}$$

Theorem A.2 (Cauchy-Peano's Theorem) *Assume that $f : I \times \Omega \to \mathbb{R}^n$ satisfies the property (H_{CP}). Then for every $(t_0, x_0) \in I \times \Omega$, there is $\varepsilon > 0$ such that the Cauchy problem (A.1) admits a solution on $[t_0 - \varepsilon, t_0 + \varepsilon]$.*

Remark A.1 The Cauchy-Peano is only an existence result. In the autonomous case, it says that if $f : \Omega \to \mathbb{R}^n$ is continuous then for every $x_0 \in \Omega$, the Cauchy problem

$$\dot{x}(t) = f(x(t)), \quad x(0) = x_0,$$

admits at least one solution locally. A counterexample to uniqueness is for example given by $f : \mathbb{R} \to \mathbb{R}$ defined by

$$f(x) := \sqrt{|x|} \quad \forall x \in \mathbb{R}.$$

The Cauchy problem $\dot{x}(t) = f(x(t))$, $x(0) = 0$ admits two smooth solutions:

$$x(t) = 0 \quad \text{and} \quad x(t) = \frac{t^2}{4} \quad \forall t \in \mathbb{R}.$$

Let $I \subset \mathbb{R}$ be an open interval, Ω be an open subset of $\mathbb{R}^n$, and $f : I \times \Omega \to \mathbb{R}^n$ be a function satisfying the following property:

(H_{CC}) For every $x \in \Omega$, there exist $\delta > 0$ and a locally integrable function $c : I \to [0, +\infty)$ such that

$$|f(t, y) - f(t, z)| \le c(t)|y - z| \quad \text{and} \quad |f(t, y)| \le c(t)$$

for almost every $t \in I$ and all $y, z \in B(x, \delta)$.

The following result provides existence and uniqueness for the Cauchy problem (A.1).

Theorem A.3 (Cauchy-Carathéodory's Theorem) *Assume that $f : I \times \Omega \to \mathbb{R}^n$ satisfies the property (H_{CC}). Then for every $(t_0, x_0) \in I \times \Omega$, there is $\varepsilon > 0$ such that the Cauchy problem (A.1) admits a solution $x : [t_0 - \varepsilon, t_0 + \varepsilon] \to \Omega$. If $y : [t_0, t_0 + \varepsilon] \to \Omega$ (or $y : [t_0 - \varepsilon, t_0] \to \Omega$) is an other solution of (A.1), then $x(t) = y(t)$ for all $t \in [t_0 - \varepsilon, t_0 + \varepsilon]$.*

Remark A.2 In the autonomous case, the Cauchy-Carathéodory Theorem says that if $f : \Omega \to \mathbb{R}^n$ is locally Lipschitz then for every $x_0 \in \Omega$, the Cauchy problem

$$\dot{x}(t) = f(x(t)), \quad x(0) = x_0,$$

admits a solution locally and this solution is unique.

By the Cauchy-Carathéodory Theorem, under assumption (H_{CC}), for every $(t_0, x_0) \in I \times \Omega$, the unique solution to the Cauchy problem (A.1) can be extended to a maximal interval of the form $I = (\alpha, \beta)$ with $\alpha < t_0 < \beta$ and $\alpha \in \mathbb{R} \cup \{-\infty\}$, $\beta \in \mathbb{R} \cup \{+\infty\}$. Under additional assumptions, we can sometimes insure that any solution can be extended to $\mathbb{R}$.

Theorem A.4 *Let $f : \mathbb{R} \times \mathbb{R}^m \to \mathbb{R}^n$ be a function satisfying the assumptions (H_{CC}) (with $\Omega = \mathbb{R}^n$) and such that there exist two functions K, M in $L^1_{loc}(\mathbb{R}, [0, +\infty))$ such that*

$$|f(t, x)| \le K(t)|x| + M(t) \quad a.e. \ t \in \mathbb{R} \ \forall x \in \mathbb{R}^n.$$

Then any solution of $\dot{x} = f(x(t))$ can be extended to $\mathbb{R}$.

Remark A.3 For sake of simplicity, we stated Theorem A.4 in the case of a nonautonomous function defined on $\mathbb{R} \times \mathbb{R}^n$. The same results holds for a function defined on $I \times \mathbb{R}^n$ where I is an open interval in $\mathbb{R}$. Namely, any solution to the Cauchy problem can be extended to I.

Let $I \subset \mathbb{R}$ be an interval and $A \in L^1(I; M_n(\mathbb{R}))$ be a function from I into the set of $n \times n$ matrices denoted by $M_n(\mathbb{R})$. By the above results, for every $t_0 \in I$, the Cauchy problem

$$\dot{S}(t) = A(t)S(t), \qquad \text{a.e. } t \in I, \quad S(t_0) = I_n,$$

has a unique solution which is defined on I. In the same way, the Cauchy problem

$$\dot{Y}(t) = -Y(t)A(t), \quad \text{a.e. } t \in I, \quad Y(t_0) = I_n,$$

admits a solution defined on I. Hence, the function $Z : I \to M_n(\mathbb{R})$ defined as $Z(t) := Y(t)S(t)$ for every $t \in I$, satisfies for almost every $t \in I$,

$$\begin{aligned} \dot{Z}(t) &= \dot{Y}(t)S(t) + Y(t)\dot{S}(t) \\ &= -Y(t)A(t)S(t) + Y(t)A(t)S(t) = 0. \end{aligned}$$

Since $Z(t_0) = I_n$, we deduce by uniqueness, that $Z(t) = I_n$ for every $t \in I$. This shows that the matrix $S(t)$ is invertible for every $t \in I$.

Proposition A.5 *Let $C \in L^1_{loc}(I; \mathbb{R}^n)$, $t_0 \in I$, and $\xi_0 \in \mathbb{R}^n$. The solution to the Cauchy problem*

$$\dot{\xi}(t) = A(t)\xi(t) + C(t), \quad \text{for a.e. } t \in I, \quad \xi(t_0) = \xi_0 \tag{A.2}$$

is given by

$$\xi(t) = S(t)\xi_0 + S(t)\int_{t_0}^t S(s)^{-1}C(s)ds, \quad \forall t \in I. \tag{A.3}$$

We check easily that the function given by (A.2) satifies (A.3).

Appendix B
Elements of Differential Calculus

We recall here basic facts of first order calculus in normed vector spaces and less basic facts of second order calculus. We refer the reader to textbook [2] for further details on differential calculus in normed spaces. The results of second order calculus are taken from the textbook [3].

B.1 First Order Calculus

Given two normed vector spaces $(X, \|\cdot\|_X)$ and $(Y, \|\cdot\|_Y)$, we denote by $\mathscr{L}(X, Y)$ the space of continuous linear maps from X to Y. This space is equipped with the operator norm (we denote alternatively by $T \cdot u$ or $T(u)$ the image of u by the operator T)

$$\|T\| = \sup\left\{ \|T(u)\|_Y \mid u \in X, \|u\|_X = 1 \right\}.$$

Let $(X, \|\cdot\|_X)$ and $(Y, \|\cdot\|_Y)$ be two normed vector spaces, $\mathscr{U}$ be an open subset of X and let $F : \mathscr{U} \subset X \to Y$ be a given mapping. Let $\bar{u} \in \mathscr{U}$. We say that F is differentiable at $\bar{u}$ provided there is a continuous linear map $D_{\bar{u}}F : X \to Y$ such that for every $\varepsilon > 0$, there is $\delta > 0$ such that

$$0 < \|u - \bar{u}\|_X < \delta \implies \frac{\|F(u) - F(\bar{u}) - D_{\bar{u}}F \cdot (u - \bar{u})\|_Y}{\|u - \bar{u}\|_X} < \varepsilon.$$

This property can also be written as

$$\lim_{u \to \bar{u}} \frac{\|F(u) - F(\bar{u}) - D_{\bar{u}}F \cdot (u - \bar{u})\|_Y}{\|u - \bar{u}\|_X} = 0,$$

or

$$F(u) = F(\bar{u}) + D_{\bar{u}}F \cdot (u - \bar{u}) + \|u - \bar{u}\|_X \, o(1).$$

L. Rifford, *Sub-Riemannian Geometry and Optimal Transport*,
SpringerBriefs in Mathematics, DOI: 10.1007/978-3-319-04804-8,
© The Author(s) 2014

The map F is said to be differentiable in $\mathscr{U} \subset X$ if it is differentiable at every $u \in \mathscr{U}$. The map

$$DF : \mathscr{U} \longrightarrow \mathscr{L}(X, Y)$$
$$u \longmapsto D_u F$$

is called the derivative of F. If DF is a continuous map on $\mathscr{U}$ (where $\mathscr{L}(X, Y)$ has the norm topology) we say that F is of class C^1 on $\mathscr{U}$. Finally we recall that given a function F of class C^1 on an open set $\mathscr{U} \subset X$ and a point $u \in \mathscr{U}$, the derivative $D_u F$ is called *singular* if it is not surjective and in that case u is called a *critical point*. The Inverse Function Theorem allows to obtain a local openness at first order.

Theorem B.1 (Inverse Function Theorem) *Let $\mathscr{U}$ be an open set of $\mathbb{R}^n$, $F : \mathscr{U} \to \mathbb{R}^n$ be a function of class C^1, and $x \in \mathscr{U}$ be such that $D_x F$ is not singular. Then there exists neighborhoods $U \subset \mathscr{U}$ of x and V of $F(x)$ such that $F_{|U} : U \to V$ is a C^1 diffeomorphism.*

One of its corollary, the Lagrange Multiplier Theorem, plays a major role in Chap. 2.

Theorem B.2 (Lagranges Multipliers Theorem) *Let $(X, \|\cdot\|_X)$ be a normed vector space, $\mathscr{U}$ be an open subset of X, and $E : \mathscr{U} \to \mathbb{R}^n$ and $C : \mathscr{U} \to \mathbb{R}$ two mappings of class C^1 on $\mathscr{U}$. Assume that $\bar{u} \in \mathscr{U}$ satisfies the following property:*

$$C(\bar{u}) \le C(u) \quad \text{for every} \quad u \in \mathscr{U} \quad \text{such that} \quad E(u) = E(\bar{u}).$$

Then there exist $\lambda_0 \in \mathbb{R}$ and $\lambda \in \mathbb{R}^n$ with $(\lambda_0, \lambda) \ne (0, 0)$ such that

$$\lambda^* D_{\bar{u}} E = \lambda_0 D_{\bar{u}} C.$$

Proof Define the mapping $\Phi : \mathscr{U} \subset X \to \mathbb{R} \times \mathbb{R}^n$ by

$$\Phi(u) := (C(u), E(u)), \quad \forall u \in \mathscr{U}.$$

The mapping Ψ is of class C^1 on $\mathscr{U}$. We claim that $\bar{u}$ is necessarily a critical point of Φ, that is $D_{\bar{u}} \Phi$ is singular. We argue by contradiction. If $\bar{u}$ is not a critical point, the continuous linear map $D_{\bar{u}} \Phi : X \to \mathbb{R} \times \mathbb{R}^n$ is surjective. Then there exists a linear subspace Y of X of dimension $n + 1$ such that the restriction of $D_{\bar{u}} \Phi$ to Y is an isomorphism. Let $y_1, \ldots, y_{n+1}$ be a basis of Y and $\mathscr{B}$ be an open neighborhood of 0 in $\mathbb{R}^{n+1}$ such that

$$\bar{u} + \sum_{i=1}^{n+1} \beta_i y_i \in \mathscr{U} \quad \forall \beta = (\beta_1, \ldots, \beta_{n+1}) \in \mathscr{B}.$$

The mapping

$$\hat{\Phi} : \mathscr{B} \longrightarrow \mathbb{R}^{n+1}$$
$$\beta = (\beta_1, \ldots, \beta_{n+1}) \longmapsto \Phi\left(\bar{u} + \sum_{i=1}^{n+1} \beta_i y_i\right)$$

is of class C^1 on $\mathscr{B}$ with a derivative which is invertible at $\beta = 0$. Hence, by the Inverse Function Theorem, the point $\Phi(\bar{u}) = (C(\bar{u}), E(\bar{u}))$ belongs to the interior of the image of $\hat{\Phi}(\mathscr{B})$. Thus for $\varepsilon > 0$ small enough, there is $y \in Y$ with $\bar{u} + y \in \mathscr{U}$ such that

$$\Phi(\bar{u} + y) = (C(\bar{u}) - \varepsilon, E(\bar{u})),$$

which contradicts. In consequence, $\bar{u}$ is a critical point of Φ. Hence, there exists a non-zero $n + 1$-tuple $p = (-\lambda_0, \lambda)$ (with $\lambda_0 \in \mathbb{R}$ and $\lambda \in \mathbb{R}^n$) which is orthogonal to the image of $D_{\bar{u}}\Phi$, that is such that

$$-\lambda_0 D_{\bar{u}} C + \lambda^* D_{\bar{u}} E = 0.$$

This concludes the proof.   □

B.2 Second Order Study

Let us denote by $\mathscr{L}^2(X, Y)$ the space of all continuous bilinear maps from $X \times X$ to Y. We can equip it with the operator norm

$$\|T\| = \sup\left\{ \|T(u_1, u_2)\|_Y \mid u_1, u_2 \in X, \|u_1\|_X = \|u_2\|_X = 1 \right\}.$$

Given an open set $\mathscr{U} \subset X$ and a mapping $F : \mathscr{U} \subset X \to Y$, we define

$$D^2 F := D(DF) : \mathscr{U} \subset X \longrightarrow \mathscr{L}^2(X, Y)$$

if it exists (where we identify $\mathscr{L}(X, \mathscr{L}(X, Y))$ with $\mathscr{L}^2(X, Y)$). If $D^2 F$ exists and is continuous on $\mathscr{U}$, we say that F is of class C^2 on $\mathscr{U}$. In this case, the second derivative $D_u^2 F$ is symmetric at any point, that is

$$D_u^2 F \cdot (v, w) = D_u^2 F \cdot (w, v) \quad \forall v, w \in X, \forall u \in \mathscr{U}.$$

If $F : \mathscr{U} \subset X \to Y$ is a function of class C^2 then we have for every $u \in \mathscr{U}$ the second order Taylor formula

$$F(u + h) = F(u) + D_u F(h) + \frac{1}{2} D_u^2 F \cdot (h, h) + \|h\|_X^2 o(1),$$

which means that

$$\lim_{v \to u} \frac{\left\| F(v) - F(u) - D_u F \cdot (v - u) - \frac{1}{2} D_u^2 F \cdot (v - u, v - u) \right\|_Y}{\|v - u\|_X^2} = 0.$$

By the Inverse Function Theorem, any function of class C^1 is locally open around any point with an invertible derivative. We are going to provide a second-order sufficient condition for local openness around critical points.

Let $(X, \| \cdot \|_X)$ be a normed vector space, N be a positive integer, $\mathcal{U}$ be an open subset of X and $F : \mathcal{U} \to \mathbb{R}^N$ be a mapping of class C^2 on $\mathcal{U}$. Given a critical point $u \in \mathcal{U}$, we call corank of u, the quantity

$$\operatorname{corank}_F(u) := N - \dim \left(\operatorname{Im}(D_u F) \right).$$

We also recall that if $Q : X \to \mathbb{R}$ is a quadratic form (that is Q is defined by $Q(v) := B(v, v)$ with $B : X \times X \to \mathbb{R}$ a symmetric bilinear form), we define its negative index by

$$\operatorname{ind}_-(Q) := \max \left\{ \dim(L) \mid Q_{|L \setminus \{0\}} < 0 \right\},$$

where $Q_{|L \setminus \{0\}} < 0$ means

$$Q(u) < 0 \quad \forall u \in L \setminus \{0\}.$$

The following result provides a sufficient condition for local openness around a critical point at second order.

Theorem B.3 *Let $F : \mathcal{U} \to \mathbb{R}^N$ be a mapping of class C^2 in an open set $\mathcal{U} \subset X$ and $\bar{u} \in \mathcal{U}$ be a critical point of F of corank r. If*

$$\operatorname{ind}_- \left(\lambda^* \left(D_{\bar{u}}^2 F \right)_{|\operatorname{Ker}(D_{\bar{u}} F)} \right) \geq r \quad \forall \lambda \in \left(\operatorname{Im}(D_{\bar{u}} F) \right)^{\perp} \setminus \{0\}, \qquad (B.1)$$

then the mapping F is locally open at $\bar{u}$, that is the image of any neighborhood of $\bar{u}$ is an neighborhood of $F(\bar{u})$.

In the above statement, $\left(D_{\bar{u}}^2 F \right)_{|\operatorname{Ker}(D_{\bar{u}} F)}$ refers to the quadratic mapping from $\operatorname{Ker}(D_{\bar{u}} F)$ to $\mathbb{R}^N$ defined by

$$\left(D_{\bar{u}}^2 F \right)_{|\operatorname{Ker}(D_{\bar{u}} F)} (v) := D_{\bar{u}}^2 F \cdot (v, v) \quad \forall v \in \operatorname{Ker}(D_{\bar{u}} F).$$

The following result is a quantitative version of the previous theorem, it is useful in Sect. 3.4. (We denote by $B_X(\cdot, \cdot)$ the balls in X with respect to the norm $\| \cdot \|_X$.)

Theorem B.4 *Let $F : \mathcal{U} \to \mathbb{R}^N$ be a mapping of class C^2 in an open set $\mathcal{U} \subset X$ and $\bar{u} \in \mathcal{U}$ be a critical point of F of corank r. If (B.1) holds, then there exist $\bar{\varepsilon}, c \in (0, 1)$ such that for every $\varepsilon \in (0, \bar{\varepsilon})$ the following property holds: For every $u \in \mathcal{U}, z \in \mathbb{R}^N$ with*

$$\|u - \bar{u}\|_X < \varepsilon, \quad |z - F(u)| < c \varepsilon^2,$$

there are $w_1, w_2 \in X$ such that $u + w_1 + w_2 \in \mathcal{U}$,

$$z = F\big(u + w_1 + w_2\big),$$

and

$$w_1 \in Ker\big(D_u F\big), \quad \|w_1\|_X < \varepsilon, \quad \|w_2\|_X < \varepsilon^2.$$

The proof of Theorems B.2 and B.3 that we give in the next sections are taken from the Agrachev-Sachkov textbook [3] and the Agrachev-Lee article [4].

We need two preliminary lemmas.

Lemma B.5 *Let $G : \mathbb{R}^k \to \mathbb{R}^l$ be a mapping of class C^2 with $G(0) = 0$. Assume that there is*

$$\bar{v} \in Ker(D_0 G) \quad with \quad D_0^2 G \cdot (\bar{v}, \bar{v}) \in Im\big(D_0 G\big),$$

such that the linear mapping

$$w \in Ker(D_0 G) \longmapsto Proj_{\mathcal{K}}\Big[D_0^2 G \cdot (\bar{v}, w)\Big] \in \mathcal{K} \tag{B.2}$$

is surjective, where $\mathcal{K} := Im(D_0 G)^{\perp}$ and $Proj_{\mathcal{K}} : \mathbb{R}^l \to \mathcal{K}$ denotes the orthogonal projection onto $\mathcal{K}$. Then there is a sequence $\{u_i\}_i$ converging to 0 in $\mathbb{R}^k$ such that $G(u_i) = 0$ and $D_{u_i} G$ is surjective for any i.

Proof Let E a vector space in $\mathbb{R}^k$ such that $\mathbb{R}^k = E \oplus Ker(D_0 G)$. Since $D_0^2 G \cdot (\bar{v}, \bar{v})$ belongs to $Im\big(D_0 G\big)$ there is $\hat{v} \in E$ such that

$$D_0 G\big(\hat{v}\big) = -\frac{1}{2} D_0^2 G \cdot (\bar{v}, \bar{v}).$$

Define the family of mappings $\{\Phi_{\varepsilon}\}_{\varepsilon > 0} : E \times Ker(D_0 G) \to \mathbb{R}^l$ by

$$\Phi_{\varepsilon}(z, t) := \frac{1}{\varepsilon^5} G\left(\varepsilon^2 \bar{v} + \varepsilon^3 t + \varepsilon^4 \hat{v} + \varepsilon^5 z\right) \quad \forall(z, t) \in E \times Ker(D_0 G), \forall \varepsilon > 0.$$

For every $\varepsilon > 0$, Φ_{ε} is of class C^2 on $E \times Ker(D_0 G) \to \mathbb{R}^l$ and its derivative at $(z, t) = (0, 0)$ is given by

$$D_{(0,0)} \Phi_{\varepsilon}(Z, T) = D_{\varepsilon^2 \bar{v} + \varepsilon^4 \hat{v}} G(Z) + \frac{1}{\varepsilon^2} D_{\varepsilon^2 \bar{v} + \varepsilon^4 \hat{v}} G(T),$$

for any $(Z, T) \in E \times Ker(D_0 G)$. For every $(Z, T) \in E \times Ker(D_0 G)$, the first term of the right-hand side $D_{\varepsilon^2 \bar{v} + \varepsilon^4 \hat{v}} G(Z)$ tends to $D_0 G(Z)$ as ε tends to 0 and since

$$\frac{1}{\varepsilon^2} D_{\varepsilon^2 \bar{v} + \varepsilon^4 \hat{v}} G(T) = \frac{1}{\varepsilon^2} \Big[D_0 G(T) + D_0^2 G \cdot \big(\varepsilon^2 \bar{v} + \varepsilon^4 \hat{v}, T\big) + \big|\varepsilon^2 \bar{v} + \varepsilon^4 \hat{v}\big| o(1) \Big]$$

$$= \frac{1}{\varepsilon^2} \Big[D_0^2 G \cdot \big(\varepsilon^2 \bar{v} + \varepsilon^4 \hat{v}, T\big) + \big|\varepsilon^2 \bar{v} + \varepsilon^4 \hat{v}\big| o(1) \Big],$$

the second term tends to $D_0^2 G(\bar{v}, T)$ as ε tends to 0. By (B.2), the linear mapping

$$(Z, T) \in E \times \text{Ker}(D_0 G) \longmapsto D_0 G(Z) + D_0^2 G \cdot (\bar{v}, T) \in \mathbb{R}^l$$

is surjective. Then there is $\bar{\varepsilon} > 0$ such that $D_0 \Phi_\varepsilon$ is surjective for all $\varepsilon \in (0, \bar{\varepsilon})$. Therefore for every $\varepsilon \in (0, \bar{\varepsilon})$ the set

$$\Big\{ (z, t) \in E \times \text{Ker}(D_0 G) \,|\, \Phi_\varepsilon(z, t) = 0 \Big\}$$

is a submanifold of class C^2 of dimension $k - l > 0$ which contains the origin. Then there is a sequence $\{(z_i, t_i)\}_i$ converging to the origin such that $\Phi_{1/i}(z_i, t_i) = 0$ and $D_{(z_i, t_i)} \Phi_{1/i}$ is surjective for all i large enough. Thus setting

$$u_i := \frac{1}{i^2} \bar{v} + \frac{1}{i^3} t_i + \frac{1}{i^4} \hat{v} + \frac{1}{i^5} z_i \quad \forall i,$$

we get $G(u_i) = 0$ and $D_{u_i} G$ surjective for all i large enough. This proves the lemma.   □

Lemma B.6 *Let $Q : \mathbb{R}^k \to \mathbb{R}^m$ be a quadratic mapping such that*

$$\text{ind}_- \left(\lambda^* Q\right) \geq m, \quad \forall \lambda \in \left(\mathbb{R}^m\right) \setminus \{0\}. \tag{B.3}$$

Then the mapping Q has a regular zero, that is there is $v \in \mathbb{R}^k$ such that $Q(v) = 0$ and $D_v Q$ is surjective.

Proof Since Q is a quadratic mapping, there is a symmetric bilinear map $B : \mathbb{R}^k \times \mathbb{R}^k \to \mathbb{R}^m$ such that

$$Q(v) = B(v, v) \quad \forall v \in \mathbb{R}^k.$$

The kernel of Q, denoted by $\text{Ker}(Q)$ is the set of $v \in \mathbb{R}^k$ such that

$$B(v, w) = 0 \quad \forall w \in \mathbb{R}^k.$$

It is a vector subspace of $\mathbb{R}^k$. Up to considering the restriction of Q to a vector space E satisfying $E \oplus \text{Ker}(Q) = \mathbb{R}^k$, we may assume that $\text{Ker}(Q) = 0$. We now prove the result by induction on m.

In the case $m = 1$, we need to prove that there is $v \in \mathbb{R}^k$ with $Q(v) = 0$ and $D_v Q \neq 0$. By (B.3), we know that $\text{ind}_- (Q) \geq 1$ and $\text{ind}_- (-Q) \geq 1$, which means that there are two vector lines L^+, L^- in $\mathbb{R}^k$ such that $Q_{|L^+ \setminus \{0\}} < 0$ and

$Q_{|L^-\setminus\{0\}} > 0$. Then the restriction of Q to $L^+ \oplus L^-$ is a quadratic form which is sign-indefinite. Such a form has regular zeros.

Let us now prove the statement of the lemma for a fixed $m > 1$ under the assumption that it has been proven for all values less than m. So we consider a quadratic mapping $Q : \mathbb{R}^k \to \mathbb{R}^m$ satisfying (B.3) and such that $\text{Ker}(Q) = \{0\}$. We distinguish two cases:

First case: $Q^{-1}(0) \neq \{0\}$.
Take any $v \neq 0$ such that $Q(v) = 0$. If v is a regular point, then the statement of the lemma follows. Thus we assume that v is a critical point of Q. Since $D_v Q(w) = 2B(v, w)$ for all $w \in \mathbb{R}^k$ and $\text{Ker}(Q) = \{0\}$, the derivative $D_v Q : \mathbb{R}^k \to \mathbb{R}^m$ cannot be zero. Then its kernel $E = \text{Ker}(D_v Q)$ has dimension $k - r$ with $r := \text{rank}(D_v Q) \in [1, m-1]$. Set $F := \text{Im}(D_v Q)^\perp$ and define the quadratic form

$$\tilde{Q} : E \simeq \mathbb{R}^{k-r} \longrightarrow F \simeq \mathbb{R}^{m-r}$$

by

$$\tilde{Q}(w) := \text{Proj}_F\big(Q(w)\big) \quad \forall w \in E,$$

where $\text{Proj}_F : \mathbb{R}^m \to F$ denotes the orthogonal projection to F. We have for every $\lambda \in F$ and every $w \in E$,

$$\lambda^* \tilde{Q}(w) = \lambda^* Q(w).$$

We claim that $\text{ind}_-(\lambda^* Q) \geq m - r$, for every $\lambda \in F \setminus \{0\}$. As a matter of fact, by assumption, for every $\lambda \in F \setminus \{0\}$ there is a vector space L of dimension m such that $(\lambda^* Q)_{|L\setminus\{0\}} < 0$. The space $L \cap E$ has dimension at least $m - k$ as the intersection of L of dimension m and E of dimension $k - r$ in $\mathbb{R}^k$. By induction, we infer that $\tilde{Q}$ has a regular zero $\tilde{w} \in E = \text{Ker}(D_v Q)$, that is $Q(\tilde{w}) \in \text{Im}(D_v Q)$ and

$$w \in E = \text{Ker}(D_v Q) \longmapsto \text{Proj}_F\big(B((\tilde{w}, w))\big) \in F$$

is surjective. Define $F : \mathbb{R}^k \to \mathbb{R}^m$ by

$$F(u) := Q(v + u) \quad \forall u \in \mathbb{R}^k.$$

The function F is of class C^2 verifies $D_0 F = D_v Q$, $D_0^2 F = B$ and the assumptions of Lemma B.5 are satisfied with $\bar{v} = \tilde{w}$. We deduce that Q has a regular zero as well.

Second case: $Q^{-1}(0) = \{0\}$.
In fact, we are going to prove that this case cannot appear. First we claim that Q is surjective. Since Q is homogeneous ($Q(rv) = r^2 Q(v)$ for all $v \in \mathbb{R}^k$ and $r \in \mathbb{R}$), we have

$$Q(\mathbb{R}^k) = \Big\{ r Q(v) \,|\, r \geq 0, v \in \mathbb{S}^{k-1} \Big\}.$$

The set $Q(\mathbb{S}^{k-1})$ is compact, hence $Q(\mathbb{R}^k)$ is closed. Assume that $Q(\mathbb{R}^k) \neq \mathbb{R}^m$ and take $x = Q(v)$ on the boundary of $Q(\mathbb{R}^k)$. Then x is necessarily a critical point for Q. Proceeding as in the first case, we infer that $x = Q(w)$ for some non-critical point. This gives a contradiction. Then we have $Q(\mathbb{R}^k) = \mathbb{R}^m$. Consequently the mapping

$$\mathscr{Q} := \frac{Q}{|Q|} \; : \; \mathbb{S}^{k-1} \longrightarrow \mathbb{S}^{m-1}$$
$$v \longmapsto \frac{Q(v)}{|Q(v)|}$$

is surjective. By Sard's Theorem, it has a regular value x, that is $x \in \mathbb{S}^{m-1}$ such that $D_v\mathscr{Q}$ is surjective for all $v \in \mathbb{S}^{k-1}$ satisfying $\mathscr{Q}(v) = x$ for all $v \in \mathbb{S}^{k-1}$. Among the set of $v \in \mathbb{S}^{k-1}$ such that $\mathscr{Q}(v) = x$ take $\bar{v}$ for which $|Q(v)|$ is minimal, that is such that

$$Q(\bar{v}) = \bar{a}x$$

and

$$\forall a > 0, \forall v \in \mathbb{S}^{k-1}, \quad Q(v) = ax \Longrightarrow a \geq \bar{a}.$$

In other terms, if we define the smooth function $\Psi : (0, +\infty) \times \mathbb{S}^{k-1} \to \mathbb{R}^m$ as,

$$\Psi(a, v) := Q(v) - ax, \quad \forall a > 0, \; \forall v \in \mathbb{S}^{k-1},$$

then the pair $(\bar{a}, \bar{v})$ satisfies

$$\bar{a} \leq a \quad \text{for every} \quad (a, v) \in (0, +\infty) \times \mathbb{S}^{k-1} \quad \text{with } \Psi(a, v) = 0.$$

By the Lagranges Multipliers Theorem (Theorem B.2), there is $\lambda_0 \in \mathbb{R}$ and $\lambda \in \mathbb{R}^m$ with $(\lambda_0, \lambda) \neq (0, 0)$ such that

$$\lambda^* D_{\bar{v}} Q = 0 \quad \text{and} \quad -\lambda^* x = \lambda_0.$$

Note that we have for every $h \in T_{\bar{v}}\mathbb{S}^{k-1} \subset \mathbb{R}^k$, we have

$$D_{\bar{v}}\mathscr{Q}(h) = \frac{1}{|Q(\bar{v})|} D_{\bar{v}}Q(h) + [D_{\bar{v}}|Q|(h)] \, Q(\bar{v}) \tag{B.4}$$

$$= \frac{1}{\bar{a}} D_{\bar{v}}Q(h) + \bar{a} [D_{\bar{v}}|Q|(h)] \, x.$$

Consequently, if $\lambda_0 = 0$ (that is if $(\bar{a}, \bar{v})$ is a critical point of ψ), then $\lambda^* D_{\bar{v}}\mathscr{Q} = 0$ which contradicts the fact $D_{\bar{v}}\mathscr{Q}$ is surjective (because λ cannot be collinear with x by 2-homogeneity of Q). In conclusion, we can assume without loss of generality that $\lambda_0 = -1$. Since $(\bar{a}, \bar{v})$ is not a critical point of ψ, the set

$$\mathscr{C} = \left\{ (a, v) \in (0, +\infty) \times \mathbb{S}^{k-1} \, | \, \Psi(a, v) = 0 \right\}$$

is a smooth submanifold of $(0, +\infty) \times \mathbb{S}^{k-1}$ of dimension $k - m$ in a neighborhood of $(\bar{a}, \bar{v})$. Then for every $(h_a, h_v) \in \text{Ker}(D_{\bar{a}, \bar{v}}\Psi)$, which is equivalent to $h_a = 0$ and $D_{\bar{v}}Q(h_v) = 0$ with $h_v \in T_{\bar{v}}\mathbb{S}^{k-1}$, there is a smooth curve $\gamma = (\gamma_a, \gamma_v) : (-\varepsilon, \varepsilon) \to \mathscr{C}$ such that $\gamma(0) = (\bar{a}, \bar{v})$ and $\dot{\gamma}(0) = (h_a, h_v)$. Then differentiating two times the equality $\Psi(\gamma(t)) = 0$ and using that $\frac{\partial^2\Psi}{\partial a^2} = 0$ and $\lambda^* \frac{\partial\Psi}{\partial v}(\bar{a}, \bar{v}) = \lambda^* D_{\bar{v}}Q = 0$, we get

$$\lambda^* \frac{\partial^2\Psi}{\partial v^2}(\bar{a}, \bar{v}) = \lambda^* \ddot{\gamma}(0)\frac{\partial\Psi}{\partial a}(\bar{a}, \bar{v}) = \ddot{\gamma}(0)\lambda^* x = \ddot{\gamma}(0).$$

Note that $\frac{\partial^2\Psi}{\partial v^2} = Q$. Furthermore, since $(\bar{a}, \bar{v})$ is solution to our minimization problem with constraine, we have $\gamma_a(t) \geq \bar{a} = \gamma_a(0)$ for all $t \in (-\varepsilon, \varepsilon)$. Then we have

$$\lambda^* Q(h) \geq 0 \quad \forall h \in \text{Ker}(D_{\bar{v}}Q) \cap T_{\bar{v}}\mathbb{S}^{k-1}.$$

Since $Q(\bar{v}) = \bar{a} > 0$ we have indeed

$$\lambda^* Q(h) \geq 0 \quad \forall h \in \left(\text{Ker}(D_{\bar{v}}Q) \cap T_{\bar{v}}\mathbb{S}^{k-1}\right) \oplus \mathbb{R}\bar{v} =: L.$$

Let us compute the dimension of the non-negative subspace L of the quadratic form $\lambda^* Q$. Since $D_{\bar{v}}\mathscr{Q}$ is surjective, we have

$$\dim\left(\text{Im}(D_{\bar{v}}\mathscr{Q})\right) = m - 1.$$

Which means (remember (B.4)) that $\text{Im}\left(D_{\bar{v}}Q_{|\mathbb{S}^{k-1}}\right)$ has dimension m or $m - 1$. But $\lambda^* D_{\bar{v}}Q = 0$ with $\lambda \neq 0$, thus we have necessarily

$$\dim\left(\text{Im}(D_{\bar{v}}Q_{|\mathbb{S}^{k-1}})\right) = m - 1$$

and

$$\dim\left(\text{Ker}(D_{\bar{v}}Q) \cap T_{\bar{v}}\mathbb{S}^{k-1}\right) = \dim\left(\text{Ker}(D_{\bar{v}}Q_{|\mathbb{S}^{k-1}})\right) = k - 1 - (m - 1)$$
$$= k - m.$$

Consequently, $\dim(L) = k - m + 1$, thus $\text{ind}_-(\lambda^* Q)$ has to be $\leq m - 1$, which contradicts the hypothesis of the lemma. This shows that $Q^{-1}(0) = \{0\}$ is impossible and concludes the proof of the lemma. $\qquad\square$

We are ready to prove Theorem B.3. Set

$$S := \left\{\lambda \in \left(\text{Im}(D_{\bar{u}}F)\right)^{\perp} \mid |\lambda| = 1\right\} \subset \mathbb{R}^N.$$

By assumption (B.1), for every $\lambda \in S$, there is a subspace $E_\lambda \subset \mathrm{Ker}\,(D_{\bar{u}} F)$ of dimension r such that

$$\lambda^* \left(D_{\bar{u}}^2 F \right)_{|E_\lambda \setminus \{0\}} < 0.$$

By continuity of the mapping $\nu \mapsto \nu^* \left(D_{\bar{u}}^2 F \right)_{|E_\lambda}$, there is an open set $\mathscr{O}_\lambda \subset S$ such that

$$\nu^* \left(D_{\bar{u}}^2 F \right)_{|E_\lambda \setminus \{0\}} < 0 \quad \forall \nu \in \mathscr{O}_\lambda.$$

Choose a finite covering

$$S = \bigcup_{i=1}^{I} \mathscr{O}_{\lambda_i}$$

and a finite dimensional space $E \subset X$ such that

$$\mathrm{Im}\left(D_{\bar{u}} F_{|E} \right) = \mathrm{Im}\left(D_{\bar{u}} F \right).$$

Then the restriction $\tilde{F}$ of F to the finite dimensional subspace $E + \sum_{i=1}^{I} E_{\lambda_i} \subset X$ satisfies

$$\mathrm{ind}_- \left(\lambda^* \left(D_{\bar{u}}^2 \tilde{F} \right)_{|\mathrm{Ker}(D_{\bar{u}} \tilde{F})} \right) \geq r \quad \forall \lambda \in \left(\mathrm{Im}(D_{\bar{u}} \tilde{F}) \right)^\perp \setminus \{0\},$$

with

$$r = \mathrm{corank}_F\left(\bar{u} \right) := N - \dim\left(\mathrm{Im}(D_{\bar{u}} F) \right) = N - \dim\left(\mathrm{Im}(D_{\bar{u}} \tilde{F}) \right).$$

Set $\mathscr{K} := \left(\mathrm{Im}\left(D_{\bar{u}} \tilde{F} \right) \right)^\perp$ and define the quadratic mapping $Q : \mathrm{Ker}(D_{\bar{u}} \tilde{F}) \to \mathscr{K}$ by

$$Q(v) := \mathrm{Proj}_{\mathscr{K}} \left[\left(D_{\bar{u}}^2 \tilde{F} \right) \cdot (v, v) \right] \quad \forall v \in \mathrm{Ker}\left(D_{\bar{u}} \tilde{F} \right),$$

where $\mathrm{Proj}_{\mathscr{K}} : \mathbb{R}^N \to \mathscr{K}$ denotes the orthogonal projection onto $\mathscr{K}$. The assumption (B.3) of Lemma B.6 is satisfied. Then by Lemmas B.6, Q has a regular zero, that is $\bar{v} \in \mathrm{Ker}(D_{\bar{u}} \tilde{F})$ such that

$$Q(\bar{v}) = 0 \quad \Longleftrightarrow \quad D_{\bar{u}}^2 \tilde{F} \cdot (\bar{v}, \bar{v}) \in \mathscr{K} = \mathrm{Im}\left(D_{\bar{u}} \tilde{F} \right)$$

and

$D_{\bar{v}} Q$ surjective

$$\Longleftrightarrow \quad w \in \mathrm{Ker}(D_{\bar{u}} \tilde{F}) \mapsto \mathrm{Proj}_{\mathscr{K}} \left[D_{\bar{u}}^2 \tilde{F} \cdot (\bar{v}, w) \right] \in \mathscr{K} \text{ surjective.}$$

Setting $G(v) := \tilde{F}(\bar{u} + v) - \tilde{F}(\bar{u})$ and applying Lemma B.5, we get a sequence $\{u_i\}_i$ converging to $\bar{u}$ such that $F(u_i) = F(\bar{u})$ and $D_{u_i}\tilde{F}$ is surjective for any i. By the Inverse Function Theorem, this implies that F is locally open at $\bar{u}$.

Proceeding as in the proof of Theorem B.3, we may assume that X is finite dimensional. We may also assume that $\bar{u} = 0$ and $F(\bar{u}) = 0$. As before, set $\mathscr{K} := (\mathrm{Im}\,(D_{\bar{u}}F))^{\perp}$ and define the quadratic mapping $Q : \mathrm{Ker}(D_0F) \to \mathscr{K}$ by

$$Q(v) := \mathrm{Proj}_{\mathscr{K}}\left[\left(D_0^2 F\right) \cdot (v, v)\right] \quad \forall v \in \mathrm{Ker}\,(D_0 F),$$

where $\mathrm{Proj}_{\mathscr{K}} : \mathbb{R}^N \to \mathscr{K}$ denotes the orthogonal projection onto $\mathscr{K}$. By (B.1) and Lemmas B.6, Q has a regular zero $\bar{v} \in \mathrm{Ker}(D_0F)$. Let E be a vector space in $\mathbb{R}^k$ such that $X = E \oplus \mathrm{Ker}(D_0F)$. Define $G : E \times \mathrm{Ker}(D_0F) \to \mathbb{R}^N$ by

$$G(z, t) := D_0 F(z) + \frac{1}{2}\left(D_0^2 F\right) \cdot (t, t) \quad \forall (z, t) \in E \times \mathrm{Ker}(D_0 F).$$

Then assumptions of Lemma B.5 are satisfied and there is a sequence $\{(z_i, t_i)\}_i$ converging to 0 such that $G(z_i, t_i) = 0$ and $D_{(z_i, t_i)}G$ is surjective for all i.

Lemma B.7 *There are* $\mu, c > 0$ *such that the image of any continuous mapping* $\tilde{G} : B(0, 1) \to \mathbb{R}^N$ *with*

$$\sup\left\{\left|\tilde{G}(u) - G(u)\right| \mid u = (z, t) \in B_X(0, 1)\right\} \leq \mu \qquad (B.5)$$

contains the ball $\bar{B}(0, c)$.

Proof This is a consequence of the Brouwer Theorem which asserts that any continuous mapping from $\bar{B}(0, 1) \subset \mathbb{R}^n$ into itself has a fixed point, see [5]. Let i large enough such that $u_i := (t_i, z_i)$ belongs to $B(0, 1/4)$. Since $D_{u_i}G$ is surjective, there is a affine space V of dimension N which contains u_i and such that $D_{u_i}G_{|V}$ is invertible. Then by the Inverse Function Theorem, there is a open ball $\mathscr{B} = B_X(u_i, \rho) \cap V$ of u_i in V such that the mapping

$$G_{|V} : \mathscr{B} \longrightarrow G_{|V}(\mathscr{B}) \subset \mathbb{R}^N$$

is a smooth diffeomophism. We denote by $\mathscr{G} : G_{|V}(\mathscr{B}) \to \mathscr{B}$ its inverse. The set $G_{|V}(\mathscr{B})$ contains some closed ball $\bar{B}(0, c)$. Taking $c > 0$ sufficiently small we may assume that

$$\mathscr{G}(y) \in B_X\left(u_i, \rho/4\right) \quad \forall y \in \bar{B}(0, c).$$

There is $\mu > 0$ such that any continuous mapping $\tilde{G} : B_X(0, 1) \to \mathbb{R}^N$ verifying (B.5) satisfies

$$\tilde{G}(u) \in G_{|V}(\mathscr{B}) \quad \forall u \in B_X(u_i, \rho/2) \cap V$$

and

$$\left|(\mathcal{G} \circ \tilde{G})(u) - u\right| \le \frac{\rho}{4} \quad \forall u \in B_X(u_i, \rho/2) \cap V.$$

Let $\tilde{G} : B_X(0, 1) \to \mathbb{R}^N$ be a continuous mapping verifying (B.5) and $y \in \bar{B}(0, c)$ be fixed. By the above construction, the function

$$\Psi : B_X(\mathcal{G}(y), \rho/4) \longrightarrow B_X(\mathcal{G}(y), \rho/4)$$

defined by

$$\Psi(u) := u - (\mathcal{G} \circ \tilde{G})(u) + \mathcal{G}(y) \quad \forall u \in B_X(\mathcal{G}(y), \rho/4),$$

is continuous from $B_X(\mathcal{G}(y), \rho/4)$ into itself. Thus by Brouwer's Theorem, it has a fixed point, that is there is $u \in B_X(\mathcal{G}(y), \rho/4)$ such that

$$\Psi(u) = u \quad \Longleftrightarrow \quad \tilde{G}(u) = y.$$

This concludes the proof of the lemma.   □

Define the family of mappings $\{\Phi_\varepsilon\}_{\varepsilon>0} : E \times \mathrm{Ker}(D_0 F) \to \mathbb{R}^N$ by

$$\Phi_\varepsilon(z, t) := \frac{1}{\varepsilon^2} F\left(\varepsilon^2 z + \varepsilon t\right) \quad \forall(z, t) \in E \times \mathrm{Ker}(D_0 F), \ \forall \varepsilon > 0.$$

By Taylor's formula at second order for F at 0, we have

$$\Phi_\varepsilon(z, t) = G(z, t) + o(1),$$

as ε tends to 0. Then there is $\bar{\varepsilon} > 0$ (with $|(\bar{\varepsilon}^2, \bar{\varepsilon})| \le 1/2$) such that for every $\varepsilon \in (0, \bar{\varepsilon})$,

$$|\Phi_\varepsilon(z, t) - G(z, t)| \le \frac{\mu}{2} \quad \forall(z, t) \in \left(E \times \mathrm{Ker}(D_0 F)\right) \cap B(0, 1).$$

By Lemma B.7 applied to $\tilde{G} = \Phi_\varepsilon$, we infer that $\bar{B}(0, c)$ is contained in $\Phi_\varepsilon\left(B(0, 1)\right)$, which in turn implies that for every $z \in \mathbb{R}^N$ such that $|z| = |z - F(\bar{u})| < c\varepsilon^2$, there are w_1, w_2 in X such that

$$z = w_1 + w_2, \quad w_1 \in \mathrm{Ker}(D_{\bar{u}} F), \quad \|w_1\|_X < \varepsilon, \quad \|w_2\|_X < \varepsilon^2.$$

Let us now show that the above result holds uniformly for u close to $\bar{u} = 0$. Since F is C^1, the vector space $\mathrm{Ker}(D_u F)$ is transverse to E for u close to $\bar{u}$. Moreover, again by C^1 regularity, for every $\delta > 0$, there is $\nu > 0$ such that for every $u \in B_X(\bar{u}, \nu)$,

$$\mathrm{Ker}(D_u F) \cap B(0, 1) \subset \left\{y + z \in X \mid y \in \mathrm{Ker}(D_{\bar{u}} F) \cap B(0, 1), \ \|z\|_X < \delta\right\}.$$

Therefore, there is $\nu > 0$, such that for every $u \in B_X(\bar{u}, \nu)$, there is a vector space $W_u \subset X$ such that (W_u could be reduced to $\{0\}$)

$$X = E \oplus W_u \oplus \mathrm{Ker}(D_u F),$$

and there are linear mappings

$$\pi_1 : \mathrm{Ker}(D_0 F) \to W_u, \quad \pi_2 : \mathrm{Ker}(D_0 F) \to \mathrm{Ker}(D_u F)$$

such that for every $t \in \mathrm{Ker}(D_0 F)$, we have

$$t = \pi_1(t) + \pi_2(t), \quad \left|\pi_1(t)\right|_X \leq K|t|, \quad \left|\pi_1(t)\right|_X \leq K|t|,$$

for some constant $K > 0$ (which depends on $\mathrm{Ker}(D_0 F)$, E, and $\|\cdot\|_X$). Given $u \in B_X(\bar{u}, \nu)$ and $\varepsilon \in (0, \bar{\varepsilon})$ we define $\tilde{G} : (E \times \mathrm{Ker}(D_0 F)) \cap B(0, 1) \to \mathbb{R}^N$ by

$$\tilde{G}(z, t) := \frac{1}{\varepsilon^2}\left(F\left(u + \varepsilon^2 z + \varepsilon^2 \pi_1(t) + \varepsilon \pi_2(t)\right) - F(u)\right),$$

for every $(z, t) \in (E \times \mathrm{Ker}(D_0 F)) \cap B(0, 1)$. Taking ν and $\bar{\varepsilon} > 0$ smaller if necessary, by Taylor's formula for F at u at second order, by the above construction and by the fact that $D_u F$ and D_u^2 are respectively close to $D_0 F$ and $D_0^2 F$, we may assume that (B.5) is satisfied. We conclude easily. □

References

1. Hirsch, M.W., Smale, S.: Differential Equations, Dynamical Systems, and Linear Algebra. Pure and Applied Mathematics, vol. 60. Academic, New York (1974)
2. Abraham, R., Marsden, J.E., Ratiu, T.: Manifolds, tensor analysis, and applications. Global Analysis Pure and Applied: Series B, 2. Addison-Wesley Publishing Co., Reading (1983)
3. Agrachev, A.A., Sachkov, Y.L.: Control Theory from the Geometric Viewpoint. Encyclopaedia of Mathematical Sciences, vol. 87, Springer, Heidelberg (2004)
4. Agrachev, A., Lee, P.: Optimal transportation under nonholonomic constraints. Trans. Amer. Math. Soc. **361**(11), 6019–6047 (2009)
5. Bredon, G.E.: Topology and Geometry. Graduate Texts in Mathematics, vol. 139. Springer, New York (1993)

Index

L. Rifford, *Sub-Riemannian Geometry and Optimal Transport*,
SpringerBriefs in Mathematics, DOI: 10.1007/978-3-319-04804-8,
© The Author(s) 2014